EN SEP 14

CORKY WILLIAMS

CORKY WILLIAMS
COWBOY POET OF THE CARIBOO CHILCOTIN

SAGE BIRCHWATER
WITH
CORKY WILLIAMS

CAITLIN PRESS

Caitlin Press Inc.
8100 Alderwood Road,
Halfmoon Bay, BC von 1y1
www.caitlin-press.com

Edit by Rebecca Hendry.
Text design by Kathleen Fraser.
Cover design by Vici Johnstone.
Printed in Canada

Caitlin Press Inc. acknowledges financial support from the Government of Canada through the Canada Book Fund and the Canada Council for the Arts, and from the Province of British Columbia through the British Columbia Arts Council and the Book Publisher's Tax Credit.

Canada Council   Conseil des Arts
for the Arts   du Canada

BRITISH COLUMBIA
ARTS COUNCIL
An agency of the Province of British Columbia

Library and Archives Canada Cataloguing in Publication
Birchwater, Sage, author
        Corky Williams : cowboy poet of the Cariboo Chilcotin
/ Sage Birchwater ; with Corky Williams.

ISBN 978-1-927575-18-5 (pbk.)

        1. Williams, Corky, 1937–. 2. Poets, Canadian (English)—21st century—Biography. 3. Ranchers—British Columbia—Cariboo Region—Biography. 4. Williams Lake (B.C.)—Biography.
I. Williams, Corky, 1937–, author II. Williams. Title.

PS8645.I4437Z65 2013   C811'.6   C2013-905494-4

CONTENTS

Map   6

Foreword   7

The Beginning   10

Corkscrew Ranch   19

Chilcotin Winters   39

The Move Downriver   60

Where Everybody Lived   73

The Big Snow   98

Herons and Marmots and Geese, Oh My!   116

Later Years   131

Williams Lake   154

Corky Back Onstage   175

End of an Era   194

Apricot Poodle Bold   204

The Ballad of the Pine Beetle   210

Cowboys and Bushrats   212

I've Got This Habit, Rabbit   214

Gone Cowboys   216

Index   218

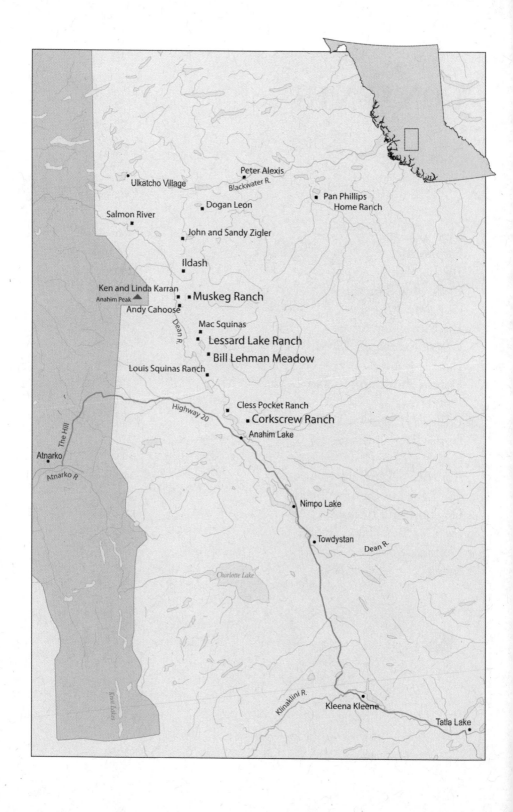

FOREWORD

I first learned of Corky Williams shortly after moving to Williams Lake in 1973. I was part of an urban collective of alternative-minded folks who established a food co-op and youth hostel on some rented property near the downtown. One day this diminutive rancher from Anahim Lake stopped by driving a mud-splattered Toyota Land Cruiser pickup truck. He was maybe five feet tall with a Texas drawl and an engaging Texas-sized personality that more than made up for any shortness in stature. He was looking for workers to help him harvest his hay crop. One of the members of our loose-knit communal group, Dana Langley, jumped at the chance for adventure and headed out with Corky to toss hay bales on his ranch, more than 200 miles west across the Chilcotin Plateau.

A month or so later Dana returned, all energized and tanned from working long hours outdoors, and he raved about this unorthodox rancher who didn't fit the typical redneck mould. He said Corky had a penchant for theatre and the arts and identified strongly with the counterculture.

In those heady days of the 1970s when anything was possible and scores of people were heading off the beaten track to find a home, I had a compulsive interest in the Chilcotin. The place felt magical. The forests went on forever and it seemed you could pick your spot and get lost for an eternity away from the busy seriousness of industrial society.

A year after Dana returned from picking bales for Corky, I bought a trapline in the mountains south of Tatla Lake, a day's journey west of Williams Lake, and began my own saga of back-woods living. This included building a homestead and gardens and raising two boys in the bush.

Corky and I would occasionally run into each other at social events, gymkhanas or rodeos, but we never really got to know each other well until 1986. That was the year of the World's Fair in Vancouver, and

Corky and I were both at significant crossroads in our lives. My relationship with the mother of my kids had ended, and I had a short-term job at the *Tribune* newspaper in Williams Lake. A year earlier Corky had been injured in a freak accident that forced him to retire from ranching, and he was in the process of re-establishing himself as an actor and performer—the work he was in before moving from Los Angeles to the Chilcotin in 1971.

Corky still has the newspaper article from May 29, 1986, that I penned about him heading off for Expo 86 to perform as a storyteller in Ian Tyson's *Cowboyography* show. He fondly recalls how they staged twenty-one sold-out performances in ten days, and how this engagement helped launch his new career in film, television and stage productions. He got an agent in Vancouver and over the next half-dozen years he did television commercials and landed parts in CBC television's *The Beachcombers* and *Lies From Lotus Land* and CTV's *Bordertown*. From there he catapulted back to Texas to join his brother Jaston Williams onstage, performing in some of the great theatres across the United States.

Meanwhile my writing career took me back to the Chilcotin, where I continued freelancing and got into authoring books. Eventually I moved to Williams Lake to take a job as a reporter for the *Williams Lake Tribune,* and one day got a phone call from Corky out of the blue. "Hey Sagebrush, how are you doing?" He had just returned to Canada after fifteen years in Texas.

Corky came right to the point. "How'd you like to help me write my story about my time in Anahim Lake?" He explained that one reason he had returned to Canada after being away all those years was to write his memoir, so we sat down to discuss how it could be done.

Ironically Corky's name had come up a couple of times in the memoir I was in the midst of completing with D'Arcy Christensen that resulted in the local bestseller *Double or Nothing: The Flying Fur Buyer of Anahim Lake* (Caitlin Press 2010). D'Arcy sold his Corkscrew Creek ranch to Corky back in 1971, then when Corky bought the Muskeg ranch further down the Dean River, as D'Arcy described it, he had a propensity to jump up on a round bale in the middle of his hayfield and moon him as he flew by in his airplane.

As we delved into Corky's story it quickly became apparent that it was more than one person's account of growing up in Texas, becoming an actor, moving to the frontier of British Columbia, then taking to the stage again. It was very much a family odyssey of endurance and survival that would be enriched by including his former wife, Jeanine, and two children, John and Dana. Fortunately Jeanine had recently moved back to Williams Lake from Texas as well, and she agreed to be part of the project.

Their story is a statement of the times, spanning the dropout generation and social unrest of the 1960s to the depopulation of British Columbia's hinterland with the passing of an old way of life in the 1980s.

Tracking the details of their adventure required several trips to the Chilcotin to meet some of the folks who were so much a part of their lives around Anahim Lake. Most notably, Bob Cohen, Big Fred Elkins, Bernie "Burnt Biscuit" Wiersbitzky, Mike Holte, Ollie Moody, Mike McDonough, Susan Hance, and Bella and Georgie Leon all shared recollections that brought the memories of those old days alive.

As it is with storytelling and the capacity to remember things that happened decades ago, some details have been forgotten. With discretion being the better part of valour, some things have been left out. Opinions shared belong to those who share them.

The Williams family odyssey depicts a slice of time in the West Chilcotin when the old ways of life were changing rapidly. When this starry-eyed Texas couple and their three-year-old son arrived, fleeing the insanity and sprawling suburbs of Los Angeles, they landed smack dab in the middle of nowhere. In Anahim Lake on the far reaches of British Columbia's Chilcotin Plateau, most people still hayed with horses and embraced a strong independent attitude, doing things the way they damn well pleased.

As Jeanine puts it, stepping into D'Arcy's store for the first time in the middle of February 1971 was like stepping back in time fifty years. "There was a wood stove with half a dozen men sitting around it and we were fascinated with them and they were pretty interested in us. We didn't realize how completely strange we seemed to them with our cowboy boots, brightly coloured nylon jackets, dark glasses and Texas accents. They couldn't have looked more amazed and amused if a pair of flamingos had set down in their midst."

—SAGE BIRCHWATER

THE BEGINNING

SAGE: It's a long way from Van Horn, Texas, to Anahim Lake, British Columbia. In 1971, it was an even longer stretch into another time and place when Corky Williams, his wife, Jeanine, and their three-year-old son, John, made the trek north via Los Angeles to begin a new life in a new country.

CORKY: Our family was farm and ranch people who had come to Texas after the Civil War. In 1947, when I was ten years old, my dad sold his property in the Texas Panhandle and we moved to Van Horn, along with my mother and my eight-year-old sister, Nina Kay. Dad bought this property because it had an ocean of good sweet water hiding underneath it that could grow anything you wanted, and nobody had dug that deep to get the water out. He was a good enough farmer to figure it out. He bought that land for fifteen dollars an acre and raised bountiful crops of grain, corn and cotton. This property was just fifteen miles from the Rio Grande and the Mexican state of Chihuahua. The Chihuahuan Desert is right there on both sides of the border.

SAGE: Corky was the elder of Jim and Vivian Williams' two surviving children when they moved to Van Horn from Lubbock, where Corky was born. The family had known tragedy. Two of Corky's older siblings had died. His sister Mary Dane died from appendicitis when she was four years old, before Corky was born, and his seventeen-year-old brother, Kenneth, was killed in a car crash when Corky was five or six. "I remember when that happened, but I was pretty young," Corky reflects. "My brother was a natural performer in school and community theatre productions, and he showed great promise. His death really hit our family hard."

CORKY WILLIAMS, COWBOY POET.

His father eventually left his government job with the State Agriculture Department and the family picked up and moved to West Texas to try a new start. When Corky was fourteen, his brother Jimmy was born, and Jimbo, as he was sometimes called, went on to become a well-known actor and playwright. He later took the stage name of Jaston Williams.

Van Horn was racially very segregated when the family arrived there in the late 1940s. There were no African-Americans living there; instead the racial divide was between whites, who lived on one side of the main street, and the people of Mexican ancestry living on the other. "The Mexican culture was in our bloodlines," Corky states. "Mexicans would come across the border and work for my dad and he treated them good. The town was segregated right down the middle; the highway ran right down the centre of Van Horn. The whites were on one side and the Mexicans were on the other. The Mexican kids went to school with us. There were signs in stores saying 'Whites Only.' Here we were going to school with Mexican kids but you couldn't invite them to your home."

Corky says most of the people where he grew up were Mexican, and he learned to speak pretty good Spanish by the time he was fifteen. "My dad put me out there with the Mexican workers, and I had to learn Spanish because they didn't speak any English. This opened up many doors for me. When I went off to college I majored in Spanish."

Corky's dad, Jim, was a pioneer around Van Horn. He raised the first bale of cotton in Culberson County, and new people wanting to move into the Van Horn area came to see him for advice on where to buy land and houses and where to build roads. "They looked to my dad as a leader in the development of the area because he was a pioneer, a great stockman and a rancher who could make the desert bloom."

Jim had a cotton farm and he would hire migrant African-American workers to pick cotton in the fall. "The Mexicans worked there too. They would work on one side of the field and the African-Americans would work on the other side. It was the poorest of the poor who were picking cotton, and my job was to weigh and record how much cotton each worker had picked. Then a guy invented a mechanical picker you could run off your tractor's power take-off, and that did away with all the labour."

Corky's mother, Vivian, was a schoolteacher, and she learned to speak Spanish too because most of her "kids," as she called them, were of Mexican descent. "She loved those kids," Corky says.

Corky first learned about Canada from his fifth-grade schoolteacher who had come from Hope, British Columbia. She had a collection of beautiful scenic photographs from all over the province, and Corky was fascinated by these images of mountains, rivers, fish and wildlife. "They looked like they were painted by the Master himself, and I told my dad, 'I'm going to live there someday.' He said, with that little grin of his, that he just might come along with me. He came to visit me many times in BC after we bought our ranch near Anahim Lake. He was stunned by the beauty of the place, and by the people who lived there."

When Corky graduated from high school, he and three friends decided to go exploring. They climbed into a car and headed north, thinking they might go to Canada, but they only got as far as Colorado.

"Where we lived in Texas was so hot, we'd never seen the mountains and snow and creeks full of water and fish, so we wanted to check it out. We came to a little town called Grand Lake, Colorado, in the high country of the Rockies. It was a beautiful place and we all got jobs working for this old guy at a place called Never Summer Ranch in those high peaks there. The government was waiting for him to die so they could take over his property as part of Rocky Mountain National Park. That summer we met a guy by the name of Dick Wright, who had a dude outfit, and the next summer we came back and worked for him, wrangling horses and taking out trail rides."

The summer season is pretty short in the mountains of Colorado because of the high elevation. "A lot of people like to go out there and go riding in the high country," Corky says. "It's a picturesque place. Rocky Mountain National Park is something else, but you have to get out of there in the early part of September because it's a mean, cold winter and it starts snowing big time."

After high school, Corky worked at various ranch jobs around Texas and hung around the rodeo scene. He tried riding a few broncs but wasn't too impressed with the hard landings, and he gave up the sport. Eventually he enrolled at Texas Tech University in Lubbock and majored in Spanish. He was a fluent Spanish speaker and had ambitions to get into a field where he could translate and write the language.

CORKY WAS BORN A PERFORMER.

After getting his degree, Corky decided to continue his education in a more practical vein and took some agriculture classes. He had room to take a couple of spare courses so he took an acting class and just loved it. Pretty soon he made performance and acting his priority and eventually graduated with a degree in theatre arts.

At Texas Tech University a friend set Corky up on a blind date with an attractive twenty-year-old sophomore from the small town of Brady in the hill country of Central Texas. Jeanine Seals was an English major with aspirations of becoming a literary agent or editor.

JEANINE: Corky and I were both attending Texas Tech University. After we met on our blind date at the Cotton Club we were quite attracted to each other. I liked Corky's wit and he was intelligent, outgoing and a good dancer. I'm a quiet person by nature, so I was quite impressed by his ability to meet people. Besides that, we both came from similar backgrounds, from small-town Texas.

CORKY: On our blind date I wore this Confederate-grey greatcoat, and Jeanine really liked that coat. Soon we got to be seeing each other every day. I'd pick her up from school in my white Cadillac Coupe de Ville and we'd go for a drive and shoot the breeze. We got to liking each other. She was very intelligent, which I really appreciated, and it evolved into a good friendship.

I had a job helping this farmer who had some Mexican workers, but he couldn't speak any Spanish. We would drive out there and I would help him communicate to his workers. Jeanine liked that because she was trying to learn Spanish.

JEANINE: I was raised in the small farming and ranching community of Brady in Central Texas. My dad worked as a station agent for the Santa Fe Railway and had a farm where we raised sheep. I was a kid who actually liked school. My mom taught me to read when I was four and I have loved books ever since. Mom had no intention of raising a rancher's wife. She wanted me to go to college and have a career. This was pretty radical thinking for those days.

I went off to college intending to get a degree in English Literature, and a job in publishing. I wanted to go to New York and be the next Max Perkins and have famous, fabulous authors and get to actually play with their work. I could take care of them in some way so they could go out and make their wonderful creations. That was what I wanted to do. How I ended up at Anahim Lake, I still don't know.

If I have learned anything in my life, it's that you can never tell what strange turns your future will take. Mine certainly hasn't followed the path I intended. When I met Corky he already had his BA in Spanish Language but had returned to school to study theatre arts. This was a pretty unusual combination for a cowboy and I was definitely impressed. I later learned that Corky's family had an artistic streak. His older brother had been showing his talent in amateur theatre productions when his life was cut short by a tragic car accident. His younger brother, Jaston, has had a very successful career as a stage actor and playwright.

We were married in 1967 and Corky finished his degree in June 1968. Corky's teachers all felt he had great talent and urged him to pursue a career in acting. Our son, John, was born in April 1968, so my education aspirations were cut short. In the summer of 1969 we moved to California. We knew that starting an acting career is often a long process so I got a job with the phone company and Corky started making the rounds of auditions.

CORKY: We ended up moving to Newhall, California, an old ranching town on the outskirts of Los Angeles. When we got there it still had a small-town flavour. One of the reasons we chose California was that my acting instructor, Dr. Clifford Ashby, knew lots of people in the film and theatre business in Los Angeles and he called them up. They said the only way to do it was to come on out, get an agent and get a

resume going. I wouldn't have gone to California because there were so
many people there, but that's what you had to do to make it.

The advice I got was to get into every play I could get in. Lots of those
agents attend those productions, so you just had to get out there and face
the firing squad. It wasn't long before I started getting quite a bit of work.

On Dr. Ashby's advice I got an agent, and it's fairly routine. They
take your photo file and resume and represent you at these auditions. I
got work doing commercials, and onstage with live theatre. I was type-
cast because of my size and my accent: it puts you in a certain category
and there weren't too many people who looked like me, so I got quite
a bit of work.

Socially, things were pretty wild in the United States at that time.
Particularly in Los Angeles. The Watts riots were really bad. Everybody
was shooting one another and they burned their own outfit down in the
ghetto and set their own neighbourhoods on fire to bring attention to
what was going on. Cops were beating the hell out of everybody for nothing. Finally the people decided they weren't going to take this anymore.

We made the decision to move to Canada, and in February 1971,
when the weather was sunny and fairly decent, we drove up to Anahim
Lake to have a look around.

JEANINE: The question people ask me the most is "Why did you move to
the Chilcotin?" There are two answers to that question. The simple one
is that I was young and stupid. The other one has to do with the way
things were in the States at that time. It was a time of social upheaval
and California was very much in the centre. There were assassinations
of leaders and riots in many cities including Los Angeles. The Vietnam
War was splitting the country apart. On a more personal level, we weren't
sure we wanted to raise our son in a place where Charles Manson was on
the loose, and more alarming yet, we found out that he and his murder-
ous family were living only a few miles away from our home. Then there
was the decapitated body of the Hells Angel found in the canyon behind
our house, and the brush fire that came within a hundred feet of our back
fence, and of course the earthquake in 1971. It really didn't seem like such
a crazy idea to pull up stakes and move to a different country.

We had a good friend who had been to the Chilcotin on a fishing
trip and he was really impressed with the country's beauty. But it was

ANAHIM LAKE, CORKY AND JEANINE'S NEW HOME.

his description of the way of life that made us seriously consider making the move. The remoteness had kept this place in a different time and that seemed to be what we really wanted. The thing that tipped the balance was the information that land and grazing rights were cheap compared to the States.

We did some hard thinking about this idea. After all, we would be moving to a foreign country, even though at that time we thought of Canada as just like the US but colder. We had a lot to learn.

We started contacting real estate agents in Williams Lake and found a place that sounded like just what we were looking for. The ranch was near the community of Anahim Lake and was owned by D'Arcy Christensen, who also owned a general store. D'Arcy's family was one of the pioneering families in the area and he was famous as the Flying Fur Buyer. We decided we'd better go up and see this place in winter, so we contacted D'Arcy and made a plan.

I will never forget that trip. The beauty of the land and the grandeur of the mountains in winter were spectacular. Every mile seemed to take us farther into another world. When we stepped into D'Arcy's store it was really like stepping back in time. It was a real general store with everything from shampoo to beaver traps. There was a wood stove

CORKY, JEANINE AND JOHN, HEADED NORTH WITH A PACKED STOCK TRAILER.

with half a dozen men sitting around it and we were introduced to the unique people who chose to live in this remote land.

We were fascinated with them and they were pretty interested in us. We didn't realize how completely strange we seemed to them, with our cowboy boots, brightly coloured nylon jackets, dark glasses and Texas accents. They couldn't have looked more amazed and amused if a pair of flamingos had set down in their midst.

Somehow we managed to ignore the warning signs: snowbanks eight feet tall, people dressed in very heavy coats and dogs with so much hair they looked like little bears. I blame D'Arcy for this. He made sure we had a great time and introduced us to some very nice people, including his family, and nobody mentioned anything about temperatures of sixty degrees below zero. It probably wouldn't have mattered by then. In the summer of 1971 we crossed the border with a stock trailer full of our possessions and headed into a new life.

CORKSCREW RANCH

JEANINE: Corkscrew Ranch was a couple of miles north of Anahim Lake. The house was of log construction and finished inside. It had three bedrooms and was very comfortable. Water for the house was pumped up from Corkscrew Creek and stored in a holding tank up in a little tower at the end of the house. There was an oil heater and the lighting was propane gas lamps.

There was a wood stove for cooking and this was my first challenge. I had never seen a real wood cookstove in my life and it took weeks of trial and error before I could turn out a half-decent meal on it. It took even longer to learn how to bake with it.

We promptly had a fireplace built and it looked great. We didn't know that we would lose more heat up the chimney than we got from the fire. This was one of our first mistakes, but it certainly wouldn't be the last. The Chilcotin is not a forgiving country. You learn or you leave.

SAGE: Bob Cohen was nineteen years old when he came to Anahim Lake from Alaska in 1956. Mentored by the likes of Lester Dorsey and Thomas Squinas, he quickly learned the ways of the country. At that time most of the Native people still lived in the bush, and Bob learned how to trap, hunt and fish from them as well and soon became very adept as a trapper, hunter and horseman.

Bob was cowboying for four ranches when Corky and Jeanine bought the Corkscrew Ranch and moved into the country as green-horns. There wasn't a more knowledgeable or capable person than Bob Cohen, and it was only natural that they hired him to cowboy their cattle and show Corky the dos and don'ts of survival in that rugged landscape.

Corky and Bob became good friends, and a few years later, when Bob took up with Francie Wilmeth, Jeanine and Francie became dear friends. Both friendships endure to this day.

CORKY: When we bought Corkscrew Ranch from D'Arcy, about two hundred head of cows came with it. Then we bought some Black Angus cows that first summer and they were the wildest things you ever did see. We turned them out on the range and never did know where they were. Bob Cohen was cowboying for us that first summer. We chased those black devils all over the country. I don't know what it was, but they were wild. They were a menace.

We took Bob's advice because he knew what to do and when to do it. He cowboyed all over that country for several ranches at the same time, and knew every creek, trail, slough meadow and mountain pasture. Bob showed me the country. I didn't even know which way to turn my horse to look it over until I got acquainted with him. He said the first thing you gotta do is learn the country. So I stayed with him all that first summer just learning the country. He told me what I needed to know: the names of the ranches, how many acres, who they belonged to or if they didn't belong to anyone.

D'Arcy had bought a red Massey-Harris square baler that went with the ranch, and we put up a mountain of square bales. At that time a lot of people were still putting up loose hay with horses. The old beaver slide stacker and old stackyards were still at Airport Meadow. You gotta have tough men to work those loose haystacks. It's an art. You have to know what you're doing to stack it so it sheds the water, and you've got to have good teams of horses. It's a labour-intensive process that could drag on for months during the summer and fall in Anahim Lake. Putting up square bales with a tractor was quicker than loose hay, but you still relied on a labour force to get the bales off the field in a hurry. If you didn't, they soaked up the moisture in those swamp meadows like sponges.

Fred Elkins, Morton Casperson, Dick Sulin and Billy Sulin worked for us too. Guys would come and work for you for three or four days, stacking hay bales mostly, and get enough money to go on a toot. Then they'd take off and there'd be another one who would take his place. Then that one would go to town and somebody else would show up.

BIG FRED ELKINS BECAME A GREAT FRIEND TO THE WILLIAMS FAMILY.

BIG FRED

SAGE: Big Fred Elkins was born in Nazko and never went to school to learn to read or write. When asked what year he was born, he fetches his First Nations status card and hands it to you. It states his birthday as June 13, 1933. Fred's mother was Southern Carrier (Dakelh) but she died when Fred was two years old. His dad, Baptiste Elkins, was Tsilhqot'in from the Chilcotin Plateau, and he left Nazko with Fred after his wife died and raised him around Anahim Lake. Fred's stepmother, Madeline Palmantier, was a half-sister to the famous Tsilhqot'in recluse Chiwid.

CORKY: I needed somebody to help me on the ranch, and D'Arcy Christensen recommended Big Fred Elkins. I ran into Lester Dorsey, and he told me the same thing. "Get Fred out there. He'll make you lots of hay. If you've got hay then you've got money. Having hay is just as good as money in the bank in Anahim Lake."

Everybody else I talked to agreed that Big Fred was the man for the job. D'Arcy sent word by the moccasin telegraph that I wanted to see Fred about a job at Corkscrew Creek. In less than twenty minutes he was at the store and we made a deal.

Big Fred was an imposing figure, around 240 pounds and the strongest man I have ever seen. I saw him jerk down a six-hundred-pound steer by the tail one time. He was bush-wise and very smart. He knew what to do and he did it.

Fred's wife, Daisy, did the cooking in the hay camp, and Fred, Morton Casperson and I did the haying. Daisy watched their kids and kept a real good camp. She made the best bannock I've ever eaten. We camped out near water and Daisy caught lots of fish, which is what we basically lived on. One of our favourite places to camp was the Louie Squinas Crossing, about five miles north of Anahim Lake on the Dean River. The bald eagles taught their young how to catch fish right at that crossing.

GEORGE CHANTYMAN WAS PRETTY CLEAR ON WHAT JOBS WERE NOT IN HIS LINE OF WORK.

I hired Fred because I needed some hands who were experienced in making hay the way they did it there in Anahim Lake. Fred was the best worker in the country. Bob Cohen once said a normal man couldn't keep up with him. Bob joked that Fred had left a string of men behind him that he had worked to death. He was just too big and strong for us to keep up with him. Building log fences, Fred would get hold of one end of those big logs and go. If a guy had hold of the other end, Fred'd drag him along as well. If Fred didn't have anything else to do he'd go out and build a log fence by himself. Bob Cohen and I were both amazed at that, because it's the hardest work in God's world.

Fred was with me all those years I was at Anahim Lake. He was our first introduction to the Native people of the area, and he soon became a big part of our lives. For all his size, he could move through the bush without making a sound. It was amazing. Fred turned out to be as great a hand as I ever had and he also became one of my best friends. He is quite a man and a fine, fine friend.

Other Native people would come by to see us, and the camp was always full of kids and dogs, especially in the summertime when the kids weren't in school. One day Fred showed me some old graves and pit houses right close to the Dean River where it flows out of Abuntlet Lake. The Dean River is fairly shallow in there, and there are a lot of fish all year. That's how people managed to survive there long ago.

About this time I noticed this little tidbit of wisdom on the bulletin board in D'Arcy's store: "Will do odd jobs or casual labor for beverage. Contact George Chanteman [Chantyman]. No shoveling horse shit please. Too much frozen." Only in the Chilcotin would you find such poetry. Everybody who read it cracked up. D'Arcy gave me the original copy and I hung it on the wall in my house. I later got to know George Chantyman very well.

Once we were finished haying Fred kept working for us. He taught us very important things we didn't know anything about.

THE LAY OF THE LAND

SAGE: The Anahim Lake/Nimpo Lake country lies in a big basin on the western edge of the four-thousand-foot Chilcotin Plateau, in the headwaters of four significant river systems. The Klinaklini River flows south into Knight Inlet, the Chilcotin River flows east into the Fraser River, the Dean River flows north, and the tributaries of Bella Coola River drain to the west.

The Dean River is the dominant feature of the landscape, as it flows northwards from Towdystan and continues to the saltchuck at Kimsquit on the Dean Channel. The country drained by the Dean River north of Anahim Lake has always been known as "downriver."

Corkscrew Creek flows into the Dean River between Little Anahim Lake and Big Anahim Lake at a place known as Goose Point. This has been an important fishing site for as long as anybody can remember. From Anahim Lake the Dean River meanders north for five or six miles past Cless Pocket Ranch into Abuntlet Lake, then narrows and picks up speed with a few rapids and falls before slowing down again, flowing gently back and forth through a rich, fertile wetland for about thirty miles to an area known as Salmon River.

In the early 1970s the country downriver was inhabited by a number of families and individuals, both Native and white, who were thoroughly engaged in living on the land. People set traplines in winter, guided hunters in the spring and fall, caught fish in the rivers and lakes, cut hay on their isolated meadows with horses to feed their small herds of livestock during the winter, worked as ranch hands, took fencing contracts or haying contracts on various ranches and moved about the country with the seasons in doing so.

The closest neighbours to Corkscrew Creek Ranch where Corky and Jeanine first settled were Thomas and Celestine Squinas, a distinguished Ulkatcho couple with nine daughters and a son. They lived on the south side of Corkscrew Creek, and Corky and Jeanine were on the north side. Jeanine says there were enough trees between the two ranches that the buildings of the two ranches weren't visible to each other. Until the 1920s, members of the Squinas clan were the only residents of Anahim Lake.

A couple of miles farther down the road was Cless Pocket Ranch, established in the late 1920s by D'Arcy's dad, Andy Christensen. With its 3,600 acres of leases and crown grants, it was by far the largest ranch in the country. Beyond Cless Pocket, the road downriver was little more than a goat trail.

JEANINE: The first place you came to past Cless Pocket Ranch was the home of Louie Squinas. He lived at the north end of Abuntlet Lake. You had to ford the Dean to get to Louie's place. His daughter Bella Leon and her husband, Georgie, lived on the other side of the river from Louie. The Holte properties, which we would eventually acquire in a land exchange with Michael Holte, were next as you continued north following the Dean River. The first was Bill Lehman Meadow at 320 acres, then Lessard Lake, Muskeg, and finally Ildash, each 160 acres. They lay in a line going down the river.

MAC SQUINAS

CORKY: Just past Lessard Lake, Mac Squinas had a few acres of Native reserve on Hump Creek. He cut a bit of hay there, but didn't stay there too much. He mostly used it as a base for his game guiding business. He helped other people hay around the country and somehow made a living, and was quite famous as a hunter. He was one of the best. He knew where things were and was one of those people who I like to say lived in the land. He was truly a part of that landscape.

Mac told me a story one time about hunting in the Rainbow Mountains. He had a pack train and a couple of American hunters and three or four guys helping him. They came around one of those switchbacks where shale rock had fallen over the trail, and they were having a hell of a time. There was no place to go except forward because the trail fell right off and it was too narrow and steep to turn around. Right at this curve, out on a ledge, Mac ran into a grizzly bear. He couldn't go back or turn his horse around because the others were right behind him. He knew if he tried to turn around he'd have stampeded all the horses off the cliff and killed them all.

Mac says he talked to that bear for a long time. He was only fifteen feet from that son of a bitch when he started talking to him. He had no other choice because he knew damn well it could be a disaster. The whole outfit could be eaten up by the bear or be killed trying to get away. "Listen, we're going to let you go by us," he told the bear. "Or you can turn around and you can go back the way you come. But I got no way to get off of this trail."

He said he talked to the bear in English and he talked to him in Carrier. "I talked to him about anything. I talked to him about a woman I used to know. I talked to that bear a long time. He knew all about my life, and when I got through talking to him he stood up, moved his head, turned around and went down the trail a little ways and hopped down that steep shale just like a dancer." Mac and the hunters stood there a long time thinking about what had just happened. "I did a lot of high-tone talking," he told me. "Maybe there is a god after all."

MORTON CASPERSON

CORKY: One day Big Fred suggested we should blow up some of the beaver dams and sandbars in Corkscrew Creek so we could turn a bit of water onto the meadow for hay production. I told D'Arcy Christensen how Corkscrew Creek was all sanded up and the water was running all over the place, and he said I should get Morton Casperson to go out there because he'd done blasting professionally. He knew what the hell he was doing. If he could blow a few of those sandbanks out of the creek, it would let the water into the meadow to irrigate the grass. Then all I'd have to do would be to pull the dams on the far end of the meadow when I wanted to drain the water and cut the hay. We asked D'Arcy to let us know the next time Morton Casperson came to town.

This swamp grass is really not a grass at all. It's a sedge that needs its roots in water to keep growing. When you drain swamps for hayland, you don't completely drain them. You get them to the right amount of standing water, and controlling the water is the main thing. Kind of like rice, it has to have lots of water. If you don't have enough water for swamp grass you get nothing.

You learn how deep to make the ditches. You don't want them too deep or they'll slough up in the low places and dry out on the high ground. Some people put a deep ditch right down the middle of the field and that's probably the worst thing you can do. You want to be real careful and make the ditches small so you can put a board in there to raise up the water level. It's simple. Pull that board out and your meadow dries so you can cut the hay, then in the springtime put the board back in there and let it flood. The key is controlling the water level with small ditches, not big ditches. Deep ditches are also a hazard to livestock and will lower the water table below the roots so the sedges won't grow.

We finally got word Morton was around and Big Fred and I ran into him in front of Baxter's store. When I first saw him he looked just like the mountain man he was reported to be. He was a living throwback in time. He had three teeth left in his head; one was white, one was yellow and one was green. He had fairly long hair and a Santa Claus beard, and he wore wide suspenders and loose-fitting woollen pants and long johns year round. He wore heavy work boots, except in spring when he changed to rubber boots like everyone else did. His hair was salt-and-pepper grey, and he had a big nicotine-stained handlebar moustache that blended into his beard. He wore an old floppy hat and always had a can of snoose in his shirt pocket, a bandana around his neck and a leather belt with a Buck knife. In the scabbard on his saddle he carried a high-powered rifle.

By the time we ran into him, Morton had heard through the moccasin telegraph that I was looking for a powder monkey to clean out some sandbars. Big Fred and I told Morton what we had in mind and he agreed to take on the job. He had everything he needed out at his place, like dynamite, fuses, caps, primer cord, crowbars, digging bars and so on. In those days you could order ditching dynamite from a hardware outfit in Prince George. You could buy it by the boxful and they'd put it on a freight truck and haul it right out to Anahim Lake like it was a sack of feed. You can't do that anymore or they'd have you under the jail.

We laid out the line of dynamite in the holes we had dug, and they all went to the main charge. This would set off all the explosives and lift the mud up and out of the sandbar in one big blast. One thing I will never forget was Morton crimping the caps onto the fuse with his teeth. He didn't have many left, but he sure could use what he had. He cut the end of the fuse with his knife so it would light easy, and when he was ready he stood up abruptly, lit the fuse and hollered, "Fire... fire...fire!" three times. The old coyote had set a three-minute fuse on the dynamite, which gave us all plenty of time to retreat from the blast area. But there was one problem. It was about that time we realized Morton's dog had not followed us.

He was just a small mutt of mixed breeding that travelled with Morton everywhere he went. Once the fuse was lit the dog was intent on checking out the area where we had set the biggest charge. There was nothing we could do to get him to follow us. We started calling him but he would not come, not even to Morton. He wanted to play, and had his rear end sticking out where we had set the dynamite. The dog was on top of a huge plug of mud and dirt when the charge went off. It was something that dog would never forget, and neither would us humans.

I thought the mutt was going to the big doghouse in the sky, but he didn't. He rode it out. He had no choice. Big Fred said, "Well, would you look at that. That dog is going as fast as I have ever seen one run. The only thing is, he can't get any traction 'cause he's twelve feet off the ground." The dog disappeared when he landed on the ground, but he finally came home after five days and survived to follow Morton for a few more years.

THE BOOZE WAS HUGE AND HORRIBLE

SAGE: Alcohol has left its mark in Chilcotin country since Western society first nudged its way into the country more than two centuries ago. Whether it was a dram of surveyors' rum, a barrel of "peaches wine" (a local expression for homebrew made from available dried fruit, sugar and yeast brewed behind the wood heater), or vanilla or lemon extract purchased by the case from the general store, people found ways to get high.

There was no liquor store in Anahim Lake when Corky and Jeanine came to the country, but people could order booze through D'Arcy Christensen's general store, or through other general stores along the highway, and have it brought in on the freight truck. Some store owners were known to keep a bottle or two of hooch squirreled away that they would trade for a beaver hide or rack of muskrat pelts.

There are many drinking stories in the country. Some are humorous, like the rancher who passed out while driving his D9 Cat down the highway and fell off. He came to his senses and saw the tracks on the road, but he wasn't sure which direction his machine had headed.

When he asked a passing tourist if she'd seen his Cat, she asked what colour it was, then tried to help by calling, "Kitty, kitty!"

Other drinking stories are more tragic, involving gunfire, homicide, maiming, traffic accidents or people staggering to their demise in sub-zero temperatures and freezing to death in a snowbank. Some people's lives were simply cut short by poor health caused by excessive boozing.

JEANINE: There was more booze in Anahim Lake country than I'd ever seen in my life. Oh my God, it was so strange to me. I grew up in a home where it was a big deal to drink a couple of beers on the Fourth of July. Most of the people I grew up with were middle-class respectable folk who might get a little tiddly at a wedding. Then we came to this country where everybody seemed to be drinking continually. I really didn't know how to handle it sometimes. Eventually I got to where I did know how, but it depended on the person. Some people got drunk and were perfectly pleasant and could be sent off to bed when the time came. Others were belligerent and unpleasant, and in that case I went to bed early and left them sitting around the kitchen table insulting each other.

There was a different kind of drinking back in those days than there is now. It was one big blowout. Somebody would go to Bella Coola and bring back a bunch of booze, there'd be a party for about a week, then everybody'd be sore and hungover and ready to go back to the country. It wasn't like today when everybody lives in town and there are three liquor stores.

Sometimes people's emotions would erupt into something that was huge and horrible. We came upon one couple, who we knew quite well, standing on the road just before Cless Pocket. We stopped because we couldn't figure out what was going on. They were having some kind of hellacious argument. Then the man started to walk away and his wife picked up a rock and wham! She hit him right in the middle of his back. She kept heaving these rocks at him. We were like, do we want to step into this or not? This is obviously a domestic dispute. Finally I guess she wore herself out. The last we saw of them they were standing there by the road still arguing.

CORKY: In that country you couldn't really get away from the booze. If you didn't have any then one of your neighbours would bring it to you.

MOONSHINE, 190-PROOF

OLLIE MOODY: The first time I ever met Corky was down at Bob Cohen's place in Anahim Lake when I brought 250 gallons of moonshine up from Oregon, hidden in a waterbed. I hit Anahim Lake five days before Stampede and was partying over at Cohen's. I left to get more moonshine, which I had stashed out, and when I come back, Corky was there. Then we got to drinking moonshine.

That moonshine was 190-proof, second run. There are gasohol machines for making fuel to run farm machinery, and this guy bought one that could make 250 gallons of gasohol at a time, but he never used it for that. He just bought it for a still so he could legally own the equipment to make moonshine. All the ingredients for 250 gallons of moonshine only cost about three hundred. I'm not sure how much I made off that moonshine, but anytime I needed money, I'd just load up ten one-gallon jugs and head off to Williams Lake. You had to keep it in glass or plastic containers, because if it's in metal, it would kill you. I'd just fill up these gallon jugs and head to the sawmill and wait in the parking lot at a certain time of the day. At lunch hour they'd meet out there and just bang, bang, bang. One hundred bucks a crack.

You could make two or three gallons from one jug of 190-proof and still have good drinking material. You take a couple shot glasses of that stuff over two hours, and you're going to have to steady yourself to stand up.

You'd run into somebody along the trail heading for town and you'd say, "Bring me four or five bottles of scotch. Anything you can pour out to get you drunk, bring it." Then they'd go to town and get the stuff and come back. They'd have four or five bottles for you. You'd pay for it. Then you did the same thing for them.

A lot of the drinking went on across from D'Arcy's store behind the log cabin where he had his salt, oats and ranch supplies. Just down the hill from there in the trees is where they had their campout. When

CORKY, HOLDING THE BOTTLE, CELEBRATES THE COMPLETION OF A
FENCING PROJECT AT MUSKEG WITH HIS WORK CREW.

they ran out, somebody would go and get some more booze, and they'd
carry on. They were passed out in the Jack pines like hogs lying in the
sunshine. You had to order your booze from town, and D'Arcy had to
make a record of it for the government. He'd sign it and that made it
legal. He tried to keep the drinking down, but once they got into it,
that was it. They were going to toot it up.

One time I was playing my mandolin and this same guy we saw arguing with his wife by the road came up to me and asked what kind of
instrument it was. I told him it was a mandolin and he said, "That's just
like the one my wife broke over my head, and a fiddle too." Another time
during an argument he escaped from town, only to be caught by his wife
in a neighbouring community where she proceeded to whoop his ass.

Years later, when we were living in the old Holte cabin at Muskeg,
Morton Casperson came out to visit and he got drunk. He got up in
the night to go outside and piss. It was dark and somehow he got his
foot stuck in our daughter Dana's potty chair. His foot got compressed
in there and he couldn't get it off. He was Norwegian from Bella Coola
and Jeanine woke me up saying, "Listen, you can hear him cussing in
Norwegian." Finally we got up and turned on the light and here was
Morton with the piss pot on his foot. He kicked the stove with it and
broke it into a thousand pieces.

Sometimes Big Fred would go on a drunk too long and I'd have a go-round with him. So he'd get sore at me and take off for a couple of months and go somewhere else to work. But when he smartened up, he'd come back.

That's the way the whole country was. You'd try and do something like get a hay crew together, build a cabin or a fence, and it was quite a challenge getting everyone pointed in the same direction. You'd get one person going, then you'd go and try and get another person. By the time you came back, your first person had wandered off to do something else, so you had to try and find them and keep your other one from wandering off. People had no sense of time. In the Chilcotin, time was just a big joke.

SAGE: Several times throughout his life Corky quit drinking, sometimes for years at a time. "It got so you couldn't go to town without somebody cramming a bottle in your face," he says. "I remember one time I went three years without having a drink. Then Roy Graham snuck some alcohol of some sort that doesn't have any taste to it into my glass. It looked like water. I don't know where he got it, but he poured some in my drink when I went to take a piss. Anyway, that got me going again. That's all it took. I was trying to quit.

"I kept trying to quit, then Bob Cohen finally did quit. It went through the country, a whole bunch of us put down the bottle. Big Fred, Ollie Moody. We started seeing people die from alcoholism. It's bad shit. God, that town was steeped in alcohol."

MEETING LINDA KARRAN

JEANINE: We'd only been living in Anahim Lake a month and were just getting settled in at Corkscrew Creek when I noticed a cabin across the creek with some people living in it. One day I was down at the creek and this woman stepped out of the cabin and came down to the creek. She was a tiny little woman, maybe ninety pounds dripping wet. She had reddish hair and really bright blue eyes. She called out to me and said hi, and we introduced ourselves. Then she said, "Why don't you come on over and we'll get acquainted?"

So I waded the creek. She said she had some dandelion wine and asked if I would like some. I'd never heard of dandelion wine but it sounded healthy and organic and mild, so sure. Well, for the next two hours we chugged dandelion wine and talked. It was one of those immediate things where you meet a person and it's so easy to talk to them you feel like you've known them forever. We talked and talked, and drank and drank. It was getting late and I knew Corky and John would be coming back fairly soon, so I thought I'd better get home and get supper on. I went to stand up and I promptly sat right back down because the whole world tilted on me. It took me a little while to realize how drunk I was.

Linda was laughing so hard she was down on the ground rolling around. I finally gathered myself up enough and Linda pointed me in the right direction. She had to help me cross the creek because I nearly fell in. I waved a happy, drunken goodbye. That was how I met Linda, and she turned out to be a great friend.

When we moved to Muskeg two years later, Linda and her husband, Ken, were our closest neighbours. They were just temporarily at the cabin across Corkscrew Creek, and moved down the Dean River to their place next to Andy Cahoose around the same time we moved to Muskeg.

CATTLEMEN'S MEETING

CORKY: A couple of months after we bought the Corkscrew place there was a meeting of the Anahim Lake Cattlemen's Association. This happened twice a year, spring and fall, and it was a social event everybody attended. It did not matter if you had cows or not; everybody came.

The meetings always took place at the community hall in the afternoon and were followed by a dance in the hall that night. The meetings were held so that ranchers could talk to the government officials about range permits, grazing leases and general business. It could get pretty heated at times, as nobody liked being told what to do with their own stock, and any regulations from the government were generally argued against pretty hard. Most of the local people regarded the government as a bunch of interfering fools who didn't know the first thing about ranching.

When I walked into my first meeting, I about fell over. Damned if Dick Wright wasn't there. At first I couldn't remember where I'd seen him before, but then he recognized me. "Aren't you Corky Williams?" he asked. Dick was the guy I worked for wrangling horses in Grand

Lake, Colorado, when I first got out of high school. He'd pioneered a piece of property up past Morrison Meadow, southeast of Anahim Lake, and was running a little sawmill there and a few head of cows. Here he was, the same guy I had worked for fifteen years earlier. I had no idea Dick had moved up there.

I noticed a bottle in a sack being passed around the room. People took a swig and passed it along. I was sitting in the back row taking this all in and pretty soon I saw another bottle making the rounds. When the bottle got to me I saw it was a gallon jug of something so I asked the guy next to me what it was. He said, "Don't ask, just drink." So I did. If a bottle ran dry, another would take its place, and this went on throughout the whole meeting.

Gradually things at the cattlemen's meeting were beginning to heat up. It looked like a Canadian version of *Gunsmoke*. These people disliked any form of government interference, especially from the Forest Service, and they let them know in no uncertain terms. As the meeting went on the crowd began to be aware of a very strange sound. As the sound got louder and louder we all started looking around to figure out where it was coming from.

We finally discovered it was coming from behind the big wood stove used to heat the hall. A large person who had obviously had a few too many pulls on the bottle had sat down there and gone to sleep. The sound was snoring like I had never heard before. It had highs and lows, beeps and whistles and a few choking gasps thrown in. The government official who was speaking had to increase his volume till he was hollering. The crowd was laughing and the bottle was still going around. Suddenly a whole new sound was heard. It was a shrill, piercing electronic whistle that had everybody putting their hands over their ears. It turned out that Pete Vogler, one of the local ranchers, was trying to turn his hearing aid down, but he had turned it up instead, and its howling, together with the snoring and hollering, created a din like nothing you have ever heard before.

Pete banged the hearing aid on his chair a couple of times and then asked the Forestry guy, who was still on the stage but had given up trying to be heard, if he knew anything about fixing hearing aids. The man allowed as he didn't and the meeting was adjourned. Forestry beat a hasty retreat from town and everybody else went to get something to eat and clear the hall to get ready for the dance that night.

BOB COHEN AND THE TROUBLE WITH FORESTRY

SAGE: Bob Cohen has never been a fan of government. When he arrived in Anahim Lake, his fiercely independent attitude melded well with the prevailing sentiment that the laws of men were best left to the dictates of nature. Many felt that the best government was no government, and Bob's government-be-damned cynicism mirrored well the feelings of many people living in the region.

JEANINE: Bob was a big force in our lives. He knew so much and we knew so little. He helped us in so many ways. He would tell us stories of the country and of the people. That was really great because we got the history. We got knowledge of these people, some of whom we had never met. Bob always had this cynical view of things. He helped us by chopping out a little of the romanticism. He always made some comment that would bust that balloon, which was good for us. He had a strong influence on us.

BOB COHEN: I was born in Alaska before it became a state. My family was up there in 1897 before the gold rush. My grandma came up alone from Kentucky and married Mr. Geist, which is German for ghost. He was a gambler and the mayor of Nome. Wyatt Earp was his bodyguard.

If you don't know the country, your cows are everywhere. I always worked for the Anahim Lake area ranging cattle in the mountains. There were four ranches paying my wages. Then in winter they looked after their own stock and I trained horses. In the summer I helped Corky, Cless Pocket, and a couple other ranches because their hands were tied. They didn't know the country, so they had to have somebody.

I worked best alone, but at roundup time I'd get some guys to give me a hand. Every summer I took two or three hundred head out to Itcha Lake. It's beautiful country, but they don't range cattle back there anymore. Forestry fouled it.

The estimation on that range was 3,500 head. I did the grass survey of it. Then Forestry said the cows were damaging the wild onion plant, when the nearest wild onion plant is Tatla Lake. You can tell by the cow's milk if it tastes like onion. The game department then said the cattle were competing with the caribou. There's only one

BOB COHEN WAS NEVER A FAN OF GOVERNMENT.

plant they were competing over in their feed and that's a green fern about six inches high they both liked. That was the only one. That was the game department being a hindrance again. They didn't do the grass survey, I did it. I gathered all the plants that the cattle would eat. Then they said you have to go down to the five-thousand-foot line. You have to do this, have to do that. Cut back, cut back, and you're in the middle of a range that would hold three thousand head with only three hundred cows.

Everybody would get after the Forestry at those cattle meetings. That way they couldn't boss you around. There were more of us than them, and that's the way it should be. We were running the outfit, not them. What the hell did Forestry know about grass range? You don't overgraze the range and cause damage if you do it right. If you over-graze the swamp grass range, you expand it. It's not like the breaks along the Fraser River where you'll have erosion if you overgraze. It's a different type of feed on those dry sidehills. But up around Anahim Lake you want to clean the range of old bottom where swamp grass grows. You've got to overgraze. That's why we burn the range.

Traditionally, we've had horses rustling all winter around Anahim

Lake. That's what your horses are for. They go in and clean up the range in the wet places where the stock can't get. The horses clean that range up all winter, working with the cattle.

The biggest problem is the government. It doesn't matter if they are foresters, agrologists or anything. They're not prepared for this type of country. They haven't studied it and haven't lived in it.

Raising cattle in this country is never a money-making deal no matter where you're at. If you have a nice ranch you'll get by, but you're not going to make a whole lot of money. Now it takes 450 head of cattle to make a living. When Corky bought Corkscrew he had 150 head and he didn't need to have half a million dollars' worth of machinery either. These guys now, with the cattle prices coming up, make a few dollars. Then the prices go back down and they're back to square one. With all that money out on machinery to feed the cows, you're held ransom by the bank. You don't make money cattle ranching. It's a way of life.

JEANINE: Bob probably spoke for a majority of the people out there. They did not like any kind of government interference in their lives and they considered Forestry to be a hindrance if not downright communistic. They felt like they knew what they needed and Forestry was just there to screw the world up. Nobody ever made any money out there.

CORKY'S PARENTS VISIT ANAHIM LAKE

CORKY: I wanted my dad, Jim, to see our place because he was a good stockman and I wanted his advice. He was absolutely stunned by the beauty of the place and by the people who lived there. He loved it and hadn't known there was any place left in the world like it. The Native people treated him like a king. He and my mother, Vivian, came up the first fall we were there. When they got to the mountains around Hope, my dad said to my mom, "Hey, this is right where I'd want to be if I was a young man."

I think he was in awe of the water around Hope, where rivers, lakes, creeks and waterfalls splash all over the place, especially since he was from the desert country of Texas where water is so scarce and so precious. I know the abundance of water really got me too.

My parents spent a couple of weeks with us that first time. A strange

CORKY'S DAD, JIM, WAS STUNNED BY THE BEAUTY OF THE CHILCOTIN.

thing happened during their visit: my mother wandered off down to Corkscrew Creek to have a look around, and damned if we didn't have to go look for her. She started following the creek, but Corkscrew Creek winds back and forth and she got turned around. She had to take pills at a certain time every day and we were quite concerned when she didn't show up. I'm not sure who ran into her first and found her. It could have been Billy Sulin. It was the fall and the days were getting shorter, and we had been haying. By the time we found her it was getting late, so it was a relief to have her back with us and still in one piece.

CHILCOTIN WINTERS

FIRST WINTER

CORKY: Moving to Anahim Lake from California was like going to the Arctic for us. That first winter was the coldest winter we ever experienced during our time in that country, and a lot of the old-timers said it was the coldest they remembered as well. It went to minus sixty-two at Puntzi and our thermometer blew out the bottom. Harold Stuart, who owned a garage at Redstone, told me one time when he was a young man it got to seventy-four below at his place. Minus sixty was plenty cold for me. Fortunately we had good friends in that country. If there hadn't been people willing to help us, to show us how to do things and teach us the way of the land, we'd have never made it. I'm sure we would have just fled. Once we realized what we'd done and thought, *Good God, what have we bought? What in the cornbread hell was I thinking of?"* it was too late. Too much whisky or something.

I remember our first Christmas on Corkscrew Creek. The water on the creek was frozen four feet deep. This was the creek where we watered our cattle. We knew nothing about providing water for two hundred head of cattle through four feet of ice in a small frozen creek, but Big Fred knew about cutting water holes. He showed us how to build ice stairs down into the creek. Cutting water holes was an art because you have to cut the hole where the water bubbles up; otherwise, if you cut a little channel, it will freeze up right in front of your face. Fred knew where to put the water holes and how to build them so the cows didn't slip into them. You have to have enough separation between the cows so they can get in there and get a drink and get back out again. Otherwise they will pile on top of one another and break a leg.

JOHN FALLS IN THE CREEK

SAGE: During winter around Anahim Lake it gets so cold that the creeks freeze to the bottom. Slush sticks to the rocks and blocks the flow of water causing the level of the creeks to rise. Often a creek will overflow its banks and freeze, creating quite a mess. As the temperature warms up slightly the water will erode a channel under the ice back into the original stream bed, leaving shelves of thin ice high and dry. Corky and Jeanine refer to these thin sheets of ice along the river as shell ice. These ice formations are dangerous if you aren't careful because you can fall through into the fast-flowing water, especially in spring as the sun's warmth melts the snow.

JEANINE: At the end of our first winter in Anahim Lake there was shell ice on Corkscrew Creek. It's one of those creeks that rises and falls a lot throughout the winter. It was early March and Billy and Dick Sulin had been working for us feeding cattle. At lunchtime they came into the house to eat. John, who would have been about four, was outside playing and I didn't think anything of it because he stayed out more than he stayed in. He loved it. He had a dog we called Butterball who was white and fuzzy—that was John's constant companion.

We sat down and I started feeding lunch to the guys. When we finished eating we were having coffee and it was quiet. All of a sudden Dick sat up in his chair and stared off into space. I could tell he had heard something, and then I heard it too. It was a far-off barking. I realized it was Butterball and I jumped up. I knew something was wrong.

Dick and Billy leaped up and shot out the door. I had to throw on some boots, and all I could find were gumboots. There was three feet of snow out there. The barking was coming from the creek, beyond where we always went for water. Corkscrew makes a big curve around a little meadow, and that's where the barking was coming from.

At that time of year there is always a hard crust on the snow. In the morning after a hard frost at night you can walk anywhere on it, but by noon it gets quite soft as the sun warms it, and you tend to fall through. It was amazing watching Dick. He didn't walk: he glided over the snow, staying on the shady side next to the trees where the snow was more frozen, moving right across the meadow. I was floundering in the

L: CORKY AND JOHN GRAINING CATTLE.
R: JOHN WITH BUTTERBALL, THE DOG WHO SAVED HIS LIFE.

deep snow, slogging down the middle of the meadow with snow filling my gumboots. I could hardly move.

Billy was coming too, doing that same sort of thing as Dick, gliding across the top of the snow. Every once in a while they'd fall through, but they'd get back up. My heart was beating so hard.

Dick got over to the willows that grow by the creek, and I couldn't see him. I kept on floundering. Billy had gone on past me too. I could still hear the dog barking, then I stopped because I was completely exhausted and I was bogged down in the snow. I knew that they were there and I just stood there watching the willows. Then I saw Dick come out, and he had John in his arms. He started sliding back across the snow. When he got up close to me he said, "He'll be okay."

I went like spaghetti. I wanted to see him and hold him. I turned around and started ploughing back through the snow, following Dick as he took him straight to the house.

John had been playing around by the creek and had gotten onto that thin shell ice and broken through. By the grace of whatever dog god there is, Butterball had somehow grabbed his hood. He wasn't quite strong enough to pull him out of the water, but he pulled him up far enough that John was out of the current and wouldn't slip under the ice. He got him partway up onto the bank and John hung onto Butterball's coat while he barked and barked and barked until we came.

He was just a mutt. I don't think he had any distinguished breeding. We didn't bring him up with us from California. I think he might have come with the ranch at Corkscrew Creek when we took the place over.

I finally managed to get back to the house and stripped John of all his clothes and wrapped him up in a blanket and towel next to the heater. It was one of those moments when half of you wanted to shake him and say, "Why did you go to the creek? You know you're not supposed to go down there." And the other half was, "Oh my son, my sweet, sweet son. I love you, I'm so glad you're alive."

I lived in terror of that creek after that. It's a swift creek. It's not dangerous for an adult most of the time, but for a kid falling in, that swift current could take them. It's one of those incredible things where the right people and the right animal were there at the right time. Because if Dick hadn't heard that barking and sat up to listen, we wouldn't have known anything was the matter. I didn't hear anything because it was so faint. Dick was a good man.

ON DEALING WITH UNFAMILIAR CRITTERS

SAGE: Moving to the West Chilcotin from the United States, Corky and Jeanine had to get used to a whole new set of animals, wildlife, birds and critters they had never seen before. Animals like porcupines, packrats and grizzlies provided challenges they were unfamiliar with, and in a country so remote, people had to be self-reliant and inventive and employ folkways and backwoods ingenuity to solve problems.

BINGO AND THE PORCUPINE

JEANINE: We had this dog, Bingo, and one day he ran into a porcupine and got stuck pretty bad in the face and mouth with quills. The way things always worked out there whenever something like that happened, people would gather around and have a meeting and give you advice on what to do.

One of the pieces of advice that seemed far-fetched at the time was to string Bingo upside down by his hind legs from a tree. Apparently that would make the blood run to his head and he would pass out. Once he had passed out, it would be easier to pull the quills out. It sounded awful for poor Bingo, but you know, it worked. I would not have given you any kind of odds on that. But it actually did work. Somebody thought of it because getting hold of a good-sized dog that doesn't want to have those things removed...that's hell.

BINGO THE DOG SURVIVES AN ENCOUNTER WITH A PORCUPINE.

CORKY: A porcupine quill has a barb in there that makes it hard to pull out. I told our friends, "Okay, we'll give it a try." I threw a rope over a tree branch and got a half-hitch around Bingo's back legs, and took the other end of the rope and tied it to the bumper of the car. Then I started backing up with it. He was so heavy it would have been difficult to lift him up any other way. We backed up gently until he was the right distance off the ground, and after he was upside down for a short time he passed out. Then we could go in there and clip those quills and pull them out. Clipping the quills unsucks them: it breaks the air seal so they pull out a lot easier. Thank God, because the pain must have been terrific. He must have pounced on the porcupine, because he had so many quills sticking out of his face, eyebrows and mouth it looked like he was wearing a beard.

JEANINE: We laid him back down on the ground when we were finished. I was worried we might have caused some brain damage, but he was all right. He stared around for a few minutes, but he was fine. His face was all puffed up for a few days, and then he completely recovered.

PL SHOOTS THE PACKRAT

JEANINE: One time we had a packrat under the sink, and PL West decided to shoot him with his .22. Why it seemed like a good idea to him, I don't know. PL, as he was known by everybody, was an Ulkatcho man who had been raised downriver by his grandparents Charlie and Jeanie West. He was living in Anahim Lake when we got there and worked for us once in a while. We later learned that his name was actually Pierre, but the old people had a hard time pronouncing "R" in the Carrier dialect and it came out like an "L."

We'd never had any experience with packrats before, and didn't know a thing about them. We heard this really strange noise and couldn't place it at first. It sounded like something rolling. It happened two or three times. This packrat had got hold of some of John's marbles and he was rolling them across the floor. There was a drop-off into the living room and you could hear the marble rolling to the step, then dropping and bouncing and rolling again.

This was happening at night. When you got up and turned the lights on, he skittered off. We were asking everyone around town, "What do we do?" The largest rodent I'd ever had to deal with before was a mouse. D'Arcy told us to get a little squirrel trap and bait it and set it. I put this little piece of bacon on the trap and put it under the sink because that seemed to be one of the places he would hide.

One night PL West and some other people were over visiting, and the packrat got bold. I went and opened a cabinet and he leaped out into the room. Of course everybody started going after him. I wish I had taken some photos or video footage of what happened that night—it was hilarious.

Somebody grabbed a broom and was going to whack him with it, and the creature ran up and jumped into a crystal bowl that Corky had inherited from his grandmother. PL was going after him with this broom handle, and Corky was screaming, "Don't break the bowl!"

The packrat was squeaking and everybody was running around when PL went outside and grabbed his .22 and came back in. By this time I was saying, "Let the rat go! No, no, no!" But the guys were into the hunt by then, and were running this thing around the house. Finally the rat ran back into the cabinet under the sink.

PL jerks the door open and the rat's sitting there looking at him, daring him to come in. PL rears back and fires. He misses the packrat, which was four feet away, and puts a bullet right through the waterline under the sink. Water was blowing everywhere.

I sat on the floor with my thumb over the hole. Oh, I was not a happy camper. Everybody left because they probably thought I was going to take that broom to them if they didn't get out.

We eventually did trap the packrat. It was more efficient than trying to shoot him. I trapped a lot of packrats after that, using a little squirrel trap and some bacon.

THE ULKATCHO PEOPLE

JEANINE: There were still quite a few Ulkatcho, Kluskus and Nazko people living in the backcountry when we arrived. Then they all started to scatter into towns and cities where their kids could go to school. Some came down our way into Anahim Lake, and some went up the other way toward Quesnel or Vanderhoof.

CORKY: It was during the 1960s and '70s that they started moving the Ulkatcho people to the Squinas reserve at Anahim Lake. They brought them out of the bush and moved them in, and built a dorm there for the kids. Before Anahim Lake was a settlement, it was the Squinas ranch. Thomas Squinas had his place over by Corkscrew Creek, and his parents, Domas and Christine, lived on the main reserve. Of course they were both gone by the time we got there, but their descendants still occupied the main ranch house.

The government put the Carriers and Tsilhqot'in all together in Anahim Lake. We got to see that. The result of it wasn't too good. They went at one another tooth and nail. It takes a few generations for everybody to accept one another. It was the smallpox epidemic of the 1860s that set it all off. It decimated their numbers and just about did them in.

Peter Alexis was one of the last of the old-timers still living in the Blackwater. He and Minnie raised all those kids. That was a tough outfit. George Chantyman lived there with one of their daughters. When I was working for the Indian Department rotating fields with my big tractor, I went up and did Peter's place and stayed in the cabin next to George Chantyman. That's how I really got acquainted with him.

JEANINE: The old Ulkatcho people were tough. Aggie Sill was so practical, using the materials she had at hand to take care of her needs. Her granddaughter Susan Hance told us how she tore off part of her skirt to swaddle her baby (Susan's father) when he was born in the bush along the trail one winter. The old women still wore the skirts when we came

ULKATCHO VILLAGE GRAVESITES.

to Anahim Lake. Three or four layers of skirts that came to mid-calf or longer. Underneath they wore moccasin leggings or heavy stockings that came up to their knees.

Besides layers of skirts they would wear layers on top too. A blouse, then a light sweater, then a heavy sweater, then a coat over that. They'd cover their heads with two or more colourful scarves. The way they dressed really made them stand out as a culture that was different from Western society. The culture seemed strong. It was like, "We are who we are. This is us." It wasn't that they were imitating the

BIRTH OF PAT SILL

SUSAN HANCE: One time my grandparents Aggie and Thomas Sill were coming back from Kiston, the place north of Ulkatcho Village where my grandfather was born, and were heading by saddlehorse to Uskisula, where they lived and had their trapline. It was November and it was very cold and the snow was deep, and Grandma Aggie was pregnant. About halfway to Uskisula, Aggie told Thomas, "I'm going into labour." Deep snow and all that, Grandpa Thomas got off the horse and built a big fire and rigged up a tent with some poles, and Granny went under the tree and delivered her own baby, which was Dad, Patrick Sill. So the baby was born and there were no baby clothes or nothing, so Grandma ripped some cloth from her dress, wrapped the baby up and put him inside her dress next to her breast. Then she got back on her horse. Can you imagine riding a horse after delivering a baby? Away they went. She said they kept checking the baby. After the long ride back to Uskisula they made it home. The only thing that happened was that my dad froze his toes. But he was okay and didn't suffer any permanent damage.

white people. They weren't giving an inch, and they didn't need to. They were strong and knew who they were.

WILD TIMES, GOOD MEMORIES

JEANINE: We made our own fun in those days. When Corky went to his first Stampede he came home with the soles burned off his boots. We found out later that Lester had to pull him away from the campfire before his feet caught on fire. I'm afraid John Barleycorn won that round.

CORKY: Sometimes it is very hard to believe what your eyes are seeing, but at Anahim Lake anything can happen at any time. It is Murphy's Law. If it can happen, it will happen, and it did. We have numerous reliable witnesses to this story who were only half drunk.

It was customary to have a dance after the stockmen's meeting in the spring. They called it the Spring Breakup Dance for a couple of reasons. Not only was the ground breaking up after the long winter freeze, but so were marriages and personal relationships after people had been cooped up together in small cabins for five or six months. They were sick of looking at one another.

Cowboys, trappers, bushmen and ranchers came to town to blow off a bunch of steam. They saddled up their horses, hitched up their teams and wagons, loaded up the women and kids and headed to Anahim Lake for the big dance and meeting. Even the dogs trailed along. Most of the people came several days ahead of time and camped out in tents along the Dean River near the rodeo grounds. That way they had water for their livestock and corrals to keep them in.

Everyone from down the Dean River and the Blackwater country had their campfires going behind the stampede grounds along the river. The stores were full, the Frontier Inn Pub was bursting at the seams, the local gas station was pumping steady and repairing flats and motel accommodations had no vacancies.

Once the stockmen's meeting was over and the government and Forest Service people had fled for their own safety, people began to turn their thoughts to the dance. There was plenty of food and drink and everyone seemed to have a pint of booze hidden on their person. This was before the RCMP attended such gatherings—the people more or less policed themselves and took care of any problems that might arise. The Spring Breakup Stockman's Dance was basically a local people's gathering that didn't attract a lot of outsiders, unlike the Stampede rodeo crowd that needed more security.

Anyway, the band showed up around nine o'clock and took to the stage and started getting tuned up. The band was a Native group from Alexis Creek called Stanley Stump and the Chieftains. They played two ways: fast and faster. The noise level in the hall was so high you couldn't always tell what song they were playing but everybody seemed to be having a great time with no sign of trouble. When they really got going good, it was pedal to the metal. By ten o'clock the place was full, with more showing up.

The band played real hard for a couple of hours and then took a break. Everybody went outside to cool off and pass the bottle. Since it was springtime there was a lot of mud around and quite a bit got tracked into the hall, where it dried on the old wooden floor. When the band started to play again and the people started dancing, a big cloud of dust began to billow up like a West Texas sandstorm. You could not see very well and sometimes you lost sight of your partner, but not to worry, another one would emerge from the cloud. Everybody had grit in their teeth and clods in their eyes.

What Anahim Lakers called dancing was more like playing rugby than doing the two-step. It was more of an energetic free-for-all. During another break all the doors and windows were opened to let in some fresh air and settle the dust. Some folks got brooms and tried to scrape some of the mud out of the hall. It helped a bit, but the dust cloud quickly returned once everyone was dancing again.

That's when a fairly large woman wearing a pair of lime-green polyester stretch pants appeared out of the haze. The way they clung to her, it looked like those pants were going to have a blowout at any moment. I was standing next to Lester Dorsey and he said, "Them green bloomers ain't gonna hold. Well now, you know what I mean."

The woman was putting her heart into dancing, jigging up and down. Her partner, an American bear hunter, seemed a little dazed by it all, but he was game and got right into the fray. As they were going around someone accidently elbowed the bear hunter from behind and he slipped, and as he went down he grabbed at his partner to break his fall. Unfortunately he grabbed her lime-green pants and down they went to her knees.

She was unfazed and never missed a beat. She pulled up her pants and went right on dancing, oblivious to the fact that she had just mooned the whole wagon train. She grabbed her partner and they disappeared into the dancing dust cloud. Everyone gave her a standing ovation. I'll never forget her and neither will anyone else who saw the lady in the lime-green pants. She was the belle of the ball.

CORKY LAUNCHES HIS ANAHIM LAKE ACTING CAREER

JEANINE: One day the guys were sitting around the house talking about Corky's acting, and they were fascinated with the whole idea that he used to get up on the stage and perform. They were asking

CORKY'S ANTICS IN ANAHIM LAKE
WERE THE BEGINNING OF A STORIED
ACTING CAREER.

him about costumes and how they were done, and Corky told them they always had special people who made the costumes and sewed them up. Then they asked about makeup, and he explained that you had to have makeup because it helped you become a different character.

Just how it came around to Corky dressing up like a woman, I'm not sure. Roy Graham asked if Corky could dress up like a woman on the stage and fool somebody, and Corky said of course he could. They loved the idea. And then, oh boy, we all got into it. I thought it was hilarious too. Of course, I had to do all the costuming. The dress was easy. I remember pulling out articles of clothing I knew I was never going to wear again in my life. Things like pantyhose, a lovely little church hat and a wig. I had bought the wig as a joke when I was in college. The hat had a veil. I also had high-heeled shoes and white gloves. A Southern girl had to have white gloves, as you never knew when you might need them. Corky had to wear gloves because his calloused hands would have given him away in a moment. To avoid the problem of someone recognizing his voice, Corky pretended he was a deaf mute.

Apparently it went off quite well. I didn't go because John wasn't feeling well. According to what everybody said, it was a great success. He had people fooled. They were pretty well snookered, and he danced with half the town. What's interesting is that the Native people weren't fooled for a minute. They knew exactly who it was, but they didn't say a word. They sat back. Thomas Squinas said they knew right off that it wasn't a woman. Not only that, he knew it was Corky.

SAGE: Roy Graham says it took Corky two days of coaching and practice to get comfortable walking around in high heels and carrying himself as a woman. By all reports he must have mastered it.

CORKY: I must have been convincing. I got pinched on the ass forty-five times. I now know how women feel. I'd been dancing with all the bachelors and they were grabbing my ass and pinching me on my fake tits while Harold Engebretson played the fiddle. It was terrible. Horny old bastards. Roy Graham's lady got all haired up and I thought she was going to attack me.

JEANINE HAS AN ECTOPIC PREGNANCY, THEN GETS PREGNANT FOR REAL

SAGE: Shortly after John's close brush with death when he fell through the shell ice along Corkscrew Creek, Jeanine had a near medical emergency of her own. She had been feeling intense abdominal pain for some time before she went to the clinic to see Dr. Dirk van der Minne, the local medical doctor who came up to Anahim Lake once a week from Kleena Kleene.

JEANINE: I think Dr. van der Minne saved my life because I had an ectopic pregnancy and I didn't know it. I didn't even know what an ectopic pregnancy was, but I found out it's a pregnancy in the fallopian tubes. I had symptoms for a while and I went in. Dirk said, "Okay, you go home, get your bag packed and get to Williams Lake as soon as you can." He sent me to Dr. Donnelly, who operated on me the next day. Dirk diagnosed my problem with no sonograms, no MRIs, nothing, just his hands. That was it, he just used his hands and he knew what it was. If I had gone on, it could have ruptured and that would have been bad news.

SAGE: Jeanine was told after her surgery that her chances of ever conceiving another child were quite slim. Six months after her operation, however, she got pregnant again, this time where nature intended it. Over the winter of 1972–73 at Corkscrew Ranch she nurtured a healthy pregnancy.

JEANINE: Two weeks before my due date, I came to town (Williams Lake), and stayed with a friend, Betty Altmier. I had false labour caused by preeclampsia and went to the hospital. They kept me in for three days. By the time I went into labour for real, Corky and John had come in. Dana was born in Williams Lake on July 10, 1973.

CORKY'S DAD HELPS EXTRICATE COWS

SAGE: That same winter Jeanine was pregnant with Dana, Corky's dad made his second visit to Anahim Lake, this time without his wife, Vivian, who was back home in Crosbyton, Texas, teaching school.

CORKY: My dad came up by himself during our second winter at Anahim Lake. It was late February, 1973, and the weather gets quite sunny and the days really start to warm up. A terrible thing had happened at Cless Pocket Ranch around Christmas when about one hundred head of cows fell through the ice and drowned. It was a hell of a sight. You looked out there and all you could see was their horns, ears and feet sticking out through the ice at Blaney Meadow.

Dad got there in February sometime, and he and I helped Lester Dorsey get those cows out of the lake. Lester got the contract to pull the cows into the timber, and Dad drove my tractor while I shackled the legs of the dead cows with chains and he dragged them out. Lester was in charge and he told us where to stack the carcasses.

The guy who was supposed to have been looking after the cows hadn't cut water holes. Somebody came by and talked him into going off to town to celebrate Christmas and he left the cattle with no drinking water. Blaney Meadow is a bad place for unsafe ice. The weather was quite mild when the guy left, and the cows could go out and get water from the lake at the edge of the ice. But when the temperature dropped, that shut their drinking hole off. The cows milled around a bit and wandered out onto the frozen lake licking the ice, and there she went. It got so heavy the whole outfit went into the lake in one plop, and they couldn't get out.

Blaney Meadow was one of the last meadows that Cless Pocket cut each year because it was so wet. Fred had worked there and said they had the same problem with watering stock that we had at Corkscrew Creek. The water in Blaney Creek would rise and fall and spill over its

YOUNG JOHN AND HIS GRANDPA JIM WILLIAMS. THE GUYS AT ANAHIM
LAKE REALLY MADE CORKY'S DAD FEEL WELCOME.

banks as the temperature got cold and the creek anchored up with ice.
Fred says it was the worst creek for that in the whole country.

We pulled those cows on top of a big hill and stacked them all up
and built a big fire and burned them. We had to get them out of the
lake and creek before it contaminated the water, and we had to dispose
of the carcasses before the bears came out of their dens. Can you imagine what kind of bear problem we would have had with one hundred
dead cows for them to feed on? The cows were bloated up and the smell
was rank even though it was the middle of winter.

After Dad helped get those cows out of the lake, all those guys shook
his hand and really made him feel welcome. He worked like hell for an
old guy. Especially for someone used to hot weather. He got right out
there and took the cold like everybody else. It did him a lot of good.

GOING PARTNERS WITH MIKE HOLTE AND MOVING DEEPER INTO THE BACKCOUNTRY

CORKY: Not long after we bought the Corkscrew Creek Ranch we heard that the Muskeg place was for sale. I heard about it from the Holte boys, Mike, Gary and Larry. It was part of the Holte estate that included four properties: Lessard Lake, Muskeg, Ildash and Bill Lehman Meadow. The Holte boys told me they didn't want to ranch any longer. They wanted to sell the property so they could divide up the money. I didn't blame them a bit because there were no facilities and everything was pretty rundown.

I was pretty interested because it meant increasing the size of our holding and getting farther into the wilderness. My dad and I talked it over. We decided to go down and talk to the new owners of Cless Pocket Ranch, Bryce and Sherry Sager, and see if they'd be interested in buying our cows, then we could start going for a hay ranch. There was always somebody who needed hay. A guy would buy a place with cattle and wouldn't have enough hay to feed them, and some ranchers never put up enough hay. Lester Dorsey said some ranchers just waved their forks at their cows and never fed them enough.

At first we made a deal with Mike Holte to go into a partnership with him, ranching together. Mike's grandparents Andy and Ada Holte had come to the Chilcotin from Washington State in the 1920s in a covered wagon. They worked their way westward across the Chilcotin Plateau and stayed for a few years in Tatla Lake, then continued on to the Engebretson Ranch at Towdystan before Andy preempted land down the Dean River.

That old boy was really smart. He started way downriver and he took up a quarter section at a time, working his way back toward Anahim Lake. Every three or four miles he took up another quarter section until he ended up with five pieces of deeded quarter sections and grazing rights that stretched about fifteen miles down the swamp along the Dean River. He tied up all the land in 160-acre blocks with the grazing rights on the outside.

So that's what we did. We got rid of most of our herd and started cultivating the land and raising hay with Mike Holte. It was easier to bring the cattle to the feed than hauling hay out to the cattle, so people brought their cows down to our place. If they were skinny when they brought them, I always fed them good and got them through the winter in good shape.

After we had ranched together for a year, Mike Holte decided he didn't want to ranch anymore at all. He had grown up at Lessard Lake as a kid and he wanted a different life besides poverty and getting starved out. Mike kept part of the Lessard Lake place to use as a base for his guiding outfit, and we traded the Corkscrew Ranch and some cash for the rest of the Holte holdings, which included Lessard Lake, Bill Lehman Meadow, Muskeg and Ildash. It took a couple of years until the deal went through. We fenced the property and it took a lot of work. Ditching, cultivating, you name it, we had to tackle it.

JEANINE AND THE PROPANE EXPLOSION

JEANINE: The value of Anahim Lake was the community and the way everybody pulled together when they needed to. It was one of the best things about that place. The time I got blown up by the propane generator is a good example. Everybody came together to help us.

When we first arrived at Corkscrew Creek the electricity for the ranch was supplied by our own diesel generator, but it was always breaking down and having problems. We decided to buy a propane generator but we didn't run it all the time.

One day when Maurice Tuck was visiting, he and I went out to start the generator. Maurice Tuck, or Tucker as we called him, was as much a part of the country as Bob Cohen, and he became our good friend too. His aunt was Mickey Dorsey, Lester's wife.

The power shed was in a little wooden building the size of an outhouse and the generator started off a battery set up inside the shed. As it turned out, this wasn't a really safe setup because a spark from the battery could set off an explosion.

I was near the door because the generator took up most of the shed. There was always a propane smell in there, but that day the smell seemed stronger than usual. Maybe there was a loose connection or something. Maurice went into the power shed first and I was right behind him. When he pushed the starter it sparked and the propane tank blew. Fortunately for both of us, it blew us both out the door. It was like a bomb went off.

When the battery sparked, it set off the tank instead of the generator. Having the one-hundred-pound tank inside the shed wasn't a good idea. The blast turned me around so I fell face-first into a snowbank. It was February and there was still lots of snow. The snow was one thing that saved me from having terrible scars. My face was burned and the explosion burned most of my hair off. If you know where to look I still have three or four tiny scars, but I was so lucky.

I was wearing nylon or polyester, which melts when it gets hot. Fortunately, the blast blew parts of the shirt right off me. Maurice was wearing thicker wool clothing and it protected him. My hands were burned pretty badly because I threw them up to protect my eyes. I think there was a moment there when the flames shot up before the explosion, so I had time to react. They went thirty feet up in the air.

Dr. Dirk van der Minne happened to be at the clinic in Anahim Lake that day. It was the one day of the week he was in the clinic. If it wasn't for that, they would have had to call for the Medevac plane from Vancouver, and it would have taken hours to get me to Vancouver.

CORKY: John went into the building right before it blew, but fortunately Tucker put him outside. If he hadn't done that, John likely would have been seriously hurt.

The blast hit Maurice Tuck in the face and the next day I took a picture of him holding the polyester shirt Jeanine had been wearing. It was just threads, like melted bubble gum. His head was swollen up and his eyes were completely closed. His face looked like a basketball. There's no way I would have recognized him. Of course Tucker wouldn't go to the doctor. When we took Jeanine in, Dr. van der Minne gave her the choice. He said they could call in the Medevac plane, which would take a couple of hours to arrive, or he could work on her himself. "I think I can do this and you won't have bad scarring," he told us.

JEANINE: I trusted him completely. I said, "Okay, I want you to do it." At that time, the medical clinic was just a little trailer on the Ulkatcho Reserve. Sister Suzanne usually took care of everything and Dr. van der Minne would come in once a week on his rounds from his home at Kleena Kleene forty miles away. So they took me in there. I don't remember a whole lot about it. I was in shock but I do remember them hooking up the IV.

Dr. van der Minne debrided my face and picked out all the little bits of hair and cloth and wood or whatever was in there, and then picked off all the burned skin.

The funny thing was how many people were in there. I can't explain how casual things were back then. Maurice was there in the treatment room, proclaiming loudly that he was fine. I think they did talk him into putting vitamin E ointment on his face, but his face wasn't burned as deeply as mine because he had thicker skin.

I was on the table with Dr. van der Minne working on me when Joel Kudra, the Pentecostal preacher, walked in. He wanted to pray with me. So he's in there praying with me, Maurice is in there blaspheming with every word; Corky is somewhere in there; and Sister Suzanne is in there. Finally Dirk bellowed at everybody, "Out! Get out! And stay out!"

It was so typical of Anahim Lake, I can tell you. Having a crowded clinic didn't really bother anybody that much. When Native people went to the doctor they liked to have their whole family with them, so Sister Suzanne and Dirk were both used to having a bunch of observers. But I guess that time he decided there were too many observers and they were too loud.

He worked on me for about three hours, going over every inch of the skin on my face, hands and chest. He was the best doctor. After that they smeared me up with this salve and bandaged my hands so I wouldn't try and use them. Then I went to Goldie Reed's house. George and Goldie Reed had a trailer in the village, over by the Anahim Lake Trading store on the edge of the reserve.

When they let me out of the clinic I didn't know where I was going to stay. I was zonked on painkillers and I guess Goldie came in and offered to let me stay at their place. I couldn't hold the baby. Dana was just a little toddler, about seven months old. She could pull herself up with the crib bars. It took me a month or six weeks to recover so I didn't look so much like a boiled tomato.

The first thing I wanted when I came out of the haze at Goldie's was to fix my hair. It was burned off in patches and it smelled awful. I told Goldie it was bothering me, and she said, "We'll see what we can do." Her daughter, Betty, was there and they somehow set me up

THE VIRTUES OF DR. DIRK VAN DER MINNE

JEANINE: Corky had a condition where his hands would throb and go to sleep when he lay down. It happened fairly frequently. I don't know how many doctors Corky had been to about this but none of them ever diagnosed it. Corky talked to Dr. van der Minne for ten minutes and the doctor knew right away it was carpal tunnel syndrome. He sent Corky down to the hospital in New Westminster and they operated on both hands. He's never been troubled with it since.

One time I had a ganglion on my hand and I didn't know what it was. I went in and he said, "Oh it's just a ganglion. It'll go away in a couple of months. We can fix it right now." He said, "Lay your hand down on the table," and then he whacked my hand with a big book. "It'll be gone by tomorrow," he said, and it was.

Some people didn't care for him because he was very blunt and outspoken, and he didn't take crap from anybody. He knew who he was, and he knew what he knew, and he was going to tell you.

We were so incredibly fortunate to have a man who had trained in Edinburgh at one of the most famous medical schools in the world.

in a chair with towels and whatnot and cut my hair. When they were finished it was about an inch long. They had to cut it that far down to get rid of the burnt hair. Needless to say I looked a little strange with hair that short, eyebrows gone, face red as a beet, and bandages on my hands.

Corky came over and brought John. When John walked in the door he stood there and looked at me. He didn't know who I was. He drew back and I started talking to him. "It's Momma. I'm all right. I'll be okay." Then he realized who I was and came on in. He was just a little boy, not yet six years old.

Hazel Mars took Dana because I couldn't take care of her. Hazel's husband, Bob Mars, was the telephone lineman for the area, and she was the operator. They lived in a trailer out on the road to Bella Coola.

It wasn't too far out of town and the trailer was provided by the phone company. We still had the hand-crank phones in those days.

It took me a couple of days to get over the shock after the explosion, then once I was well enough to get up and go out, the first thing I wanted to do was go and see Dana. The strange thing was, she didn't show any reaction. I made sure I started talking when I saw her so she could hear my voice. I didn't want to traumatize her.

I stayed at Goldie's place for about a week, then I went home. Dana stayed with Hazel for two or three more days after I left Goldie's until I could use my hands again, then I wanted her home. Luckily the insides of my hands weren't burned as badly as the backs. I had thrown my hands up to protect my face and the backs of my hands had taken the brunt of the explosion.

John had been old enough to realize the danger when it happened. All the adults were freaking out and his mom was lying there in the snow. I'm sure he was terrified and couldn't understand or put the pieces together. His mom disappeared, and then seeing this person, this being, this thing…it must have freaked him out.

Of course, everybody knew what happened within half an hour. Other people in the community helped out as well. John was old enough to stay with Corky, and people made food for them or took them out to eat or brought them home to eat with them. I looked pretty weird for several weeks. I had to go into the clinic when Dirk came around on his weekly visits to get my face washed. It was not fun. He took a regular old washcloth and put some Isoderm on it. Whatever he was doing worked. I only had a few very small scars I could cover with makeup. Dr. van der Minne did a good job.

THE MOVE DOWNRIVER

SAGE: By the late spring of 1974, Jeanine had fully recovered from the propane tank explosion, and she and Corky and the kids headed out from Corkscrew Creek Ranch to begin their life downriver inhabiting the old Holte property at Muskeg. In dry conditions it would have meant a tedious two-hour journey on the rutted wagon road for twenty miles, bumping over the tree stumps, potholes and boulders that Corky jokingly refers to as Anahim Lake pea-gravel.

But springtime is anything but dry. The winter frost is just escaping the ground, and for good reason the locals refer to this season as spring breakup. That's when normally solid ground becomes a bottomless quagmire of mud. Exacerbating the breakup season is the moisture from the melting snow, which typically keeps the ground sodden well into summer. Not surprisingly then, this initial foray to Muskeg was an epic adventure they wouldn't soon forget, as they lurched from bog hole to bog hole.

JEANINE: We were loaded up with all my chickens and everything in the truck, and we got stuck so bad just past Cless Pocket that Bryce Sager had to pull us out with his Cat. Having Bryce and Sherry Sager running the Cless Pocket Ranch was a welcome addition to the community. After Cless Pocket we got stuck at Louie Squinas's turnoff by Abuntlet Lake, but we managed to get ourselves out of there. Then we got bogged down again, beyond any hope of moving, just past the little bridge at Betty Creek. It was fantastic good fortune that Bill Lampert happened to come along with his Cat from the Blackwater and pull us out of there. Twenty miles of road and we'd been on there for eight hours by that time. I took John and Dana and walked the last two miles from Betty Creek into Muskeg. It was misting rain and cold, and

THE OLD HOLTE CABIN AT MUSKEG.

we were all soaked by the time we got there. When I went to open the door of the cabin, it was locked. Even though there was nothing in it, for some reason someone had locked it. I went berserk. I think I scared my children. I was wearing gumboots and I reared back and kicked that door as hard as I could, and it popped the lock right off. My temper was a little bit frayed. Then I went in and started a fire and put the coffee on and had things semi-organized by the time the guys got there. We stayed at Muskeg all summer and started fixing the place up.

CORKY: We looked like the Beverly Hillbillies when we set out from Corkscrew. We even had a rocking chair up on the top of the load. We moved out of Corkscrew in stages and didn't just leave in one big move. We did a lot of back and forth that summer as we cut hay at Corkscrew, Muskeg and Ildash. There was no rush to get out of Corkscrew because all that land division hadn't gone through with the Holtes.

MUSKEG

SAGE: In the spring of 1974, as the family got settled into the old homestead cabin at Muskeg built by Andy Holte in the dirty thirties, Dana was approaching her first birthday and John was six. There was plenty of work

JEANINE AND DANA AT THE OLD
MUSKEG CABIN.

to do to get the ranch up to snuff. It was seriously rundown after years of neglect. Miles of fences had to be built or replaced. The cabin was old but comfortable, and Jeanine made the rustic structure into a cozy home despite its leaky roof and interior decorating of flattened cardboard boxes nailed to the log walls. There were cracks between the logs and around the windows that gave ample passage to cold air and bugs of every description. The mosquitoes and blackflies took some getting used to. At Corkscrew the bugs had been bad, but at Muskeg they were unbelievable. Corky and Jeanine learned to cope by draping their beds with mosquito netting, lighting mosquito coils, building smudges outside for the animals and applying generous quantities of bug dope to their skin.

Big Fred and Daisy set up camp at Muskeg that summer as well. Camping was a lifestyle they were used to and relished. Their four kids, John, Garren, Charlene and Shammy, were ready playmates for Corky and Jeanine's John and Dana, although Dana and Charlene were still babies. They became the best of friends and played together every day. Some things are ironic. Not only did both families have a son named John, but both Johns had the same middle name: Lawrence. To distinguish between the two, John Elkins was referred to as John Lawrence, and John Williams was called Little John, because he was six or seven years younger.

That summer Corky hired Pat Sill and his family to start repairing and building the fences at Muskeg. Soon other fence builders like Andy Cahoose and Dogan Leon and their families joined in, and the big job of fencing the property progressed quickly. Like Big Fred and Daisy, everyone set up camp in the timber, where they would be protected.

WELCOMED BY THE URSUS HORRIBILIS

CORKY: About a week after we arrived at Muskeg that spring, Big Fred and I were off looking at some of the swamp I wanted to drain for hay land. This part of the meadow had old grass six feet tall, and the hay

CLAYTON MACK, MARIANNA BEHM AND CORKY IN BELLA COOLA, WITH ANOTHER URSUS HORRIBILIS SPECIMEN.

crop was thick and heavy. As we pushed our way into a clearing we heard what sounded like bones being crunched and broken.

Big Fred had an old 30-30 rifle with him. You could look down the barrel and see it was obviously crooked. I had once said to Fred, "It looks to me like you've got a gun that shoots around corners." He had laughed like hell about that and said, "I may have."

Silently Big Fred parted the grass to see what was making the sound. Less than two hundred yards away, a big grizzly bear stood up on its hind feet and started sniffing the air. We were lucky the wind was blowing the right way and he didn't smell us. The bear dropped down and we could hear more bones being crunched. It was springtime and the bears just out of hibernation were hungry. They would eat anything from mice to moose, cleaning up whatever had died over the winter in the forest.

Ursus horribilis is the king of the jungle in that wild country. He is very smart and cunning, and very wily. He can outrun the fastest saddle horse, especially in downed windfall, and has no fear of anything. He likes to dig a hole in the ground to bury his prey and let it rot for a few days because he prefers to eat the rotten flesh. Carnage left over from winter is a particular delicacy. But a grizzly is irritable and will kill your ass and eat it too if disturbed from his meal.

Big Fred and I got moving fast. We silently got out of sight of the bear and made a big circle, crossing over the creek on a huge beaver dam, all the while staying upwind so the bear couldn't smell us. We were now on the opposite side of the creek from the bear and up on a gravel ridge above the meadow. We got into the timber and snuck around behind the bear until we could see him through the trees below us. He was a huge grizzly chewing on the carcass of a large animal, but we couldn't tell what kind of animal it was in the thick undergrowth.

We weren't comfortable having a hungry grizzly a quarter-mile away from where we were camped with our families, so Fred said we should shoot him. All he had was his old, bent 30-30, and I was unarmed, though after that episode I always carried a rifle too. We eased up to the edge of the timber, which gave Fred the best shot he was ever going to get. While he was waiting for the bear to turn around and face us, I whispered, "Don't shoot him until I get up in this tree."

He said, "I'll shoot it when I get a good shot."

I was only halfway up the tree when the bear stood up. Fred was standing on the ground below me, resting his rifle on a tree limb. He took deadly aim, clicked the hammer back and pulled the trigger. The gun roared, and almost instantly one hell of a roar came out of that bear too. He jumped about six feet straight up in the air and I could see where the bullet had blown a chunk of muskeg out from behind him. He wasn't hit, but the bullet might have grazed his groin going between his hind legs. The bear ran down to the creek and immediately started looking around for his attacker. This was a real bad sign.

He crossed the creek running toward us, and when he hit the water it exploded. He ran a short distance in our direction and then stopped. Fred took another bead on him with his old rifle but it wouldn't fire, so he threw the gun down, grabbed the tree I was in and started climbing. I told him to find his own damn tree. Fred was so big I was afraid the tree wouldn't hold us both. But Fred wasn't listening, so I climbed a little higher to make room.

We watched as the bear headed back across the creek and started running in big circles around the meadow. Each circle got bigger and bigger and you could hear the hollow sound of his feet pounding on the muskeg. Boom, boom, boom! He was trying to figure out which direction that shot had come from, and what direction he should run. We couldn't tell if he was on the fight or looking to retreat. It would have been difficult for the bear to know exactly where his attacker was because of the way the sound reverberates off the mountains and echoes off trees along the edge of the meadow. Fortunately, the wind was blowing from the bear to us and he couldn't smell us. Finally, after three or four big circles, he ran out of the swamp and into the trees and we never saw him again. Big Fred and I stayed in the tree for about half an hour. Finally we looked around the trunk at each other and we both laughed. We climbed down and discovered that the bear had been feeding on the carcass of a horse that had died that winter.

I said to Fred that maybe his rifle shoots around corners pretty good, but aiming straight on it wasn't so dependable. He laughed about that and said he'd take care of that problem. He sighted it in the next day and after that he rarely missed anything he aimed at, bent barrel and all.

THE TRAGIC DEATH OF MURRAY VANNOY

CORKY: About six weeks after our close brush with the grizzly bear, D'Arcy Christensen buzzed our place with his airplane. It was the third week in July and the custom we had in those days before we had CB radios was that if there was an emergency and no other way to communicate, one of the pilots would write up a message for you, put it in a container, fly over your ranch and drop it down to you.

JEANINE: We all came out to see what was going on and D'Arcy Christensen started circling. Then we saw him toss something out the window of his plane. It was a coffee can with a streamer of flagging tape tied on to it. When we picked it up in the meadow, there was a message telling us that Murray Vannoy had been killed over in the Blackwater and they needed our help. Murray and his wife and four kids lived in an isolated place between Eliguk Lake and Pan Phillips' Home Ranch. The note said his tractor had turned over on him on a trail he was building to Laidman Lake, and they couldn't get it off his body. Corky had the only four-wheel-drive tractor in the country and they wanted him to drive out there to help lift the tractor off Murray's body.

CORKY HAYING THE BRUSH MEADOW AT MUSKEG WITH HIS FOUR-WHEEL-DRIVE UNIVERSAL TRACTOR, MANUFACTURED IN ROMANIA.

That meant going in by way of Rainbow Lake, up Woman Hill and past Eliguk Lake, through bogs, slogs and the awful stump road to Laidman Lake. So I got food together while Corky and Fred loaded the wagon with all the equipment they thought they might need for the trip. They put chains on all four wheels of the tractor because it was still muddy and wet out in the bush, and they headed up there.

SAGE: Little John was only six years old but he was used to working alongside Corky and Big Fred, and he remembers being really upset when they headed out to the Blackwater without him. "I remember when Dad and Fred took off," he says. "I wanted to go, but Dad said no. There was a big pout at the gate and I was screaming and got sent home. I didn't realize it then, but it would have been a hard trip."

CORKY: It took Fred and me two or three days to get from Muskeg to where Murray had his accident—we camped the first night at Rainbow Lake with John and Sandy Zigler. We didn't know what we were facing until we got over there. We had loaded up stuff like come-alongs, chains, cables, power saws, shovels and axes. When we got to the Blackwater we met the people from Moose Lake who had a team and wagon, and I think we abandoned the tractor somewhere along the trail because it was so rough. We put all the stuff we thought we'd need on the horse-drawn wagon, and I rode in on saddle horse with Murray Vannoy's oldest boy, Randy. We were off by ourselves and I talked to him all the way there. He was a Christian kid and was handling it well. He had a lot of faith.

He wouldn't say much at first because he was so bushed. I got to talking to him about religion and people passing on and stuff, and that brought him out a little bit.

I found out that Murray had sent his wife and three younger children ahead with their team and wagon to wait for him at a certain point. Randy had stayed with his dad and saw the whole thing happen. The kid said when the tractor turned over and pinned his dad to the ground, he had just enough air to tell Randy to go catch up with his mother because he wasn't going to make it.

It shook me up to have the boy there, even though he told me he was all right and had been there when the accident happened. I told him I didn't want him to go in and see his dad until I went in there and had a look. He said his dad's spirit had already gone. "It's just the body," he said. "He's gone to a better place." Here's a nine-year-old boy telling me this and his father's dead with a tractor laying on top of him. He was pretty cool, let me tell you. He took it good. He never cried. I have to give him credit for a kid who had just lost his father.

We were the only ones there at first, me and the boy. Then eventually everyone else got there with the team and wagon and a couple of other people on horseback. To get the tractor off Murray's body, we ran cables up to a tree then used a block and tackle to pull the tractor forward. Murray's hands were still frozen to the steering wheel with rigor mortis, and he rode the tractor up and out of the mud. He rose up like a phantom. It was wild looking, let me tell you. His body was blue, and it took me a bit to get him off of there. I had to pry his fingers off the steering wheel.

The seats on those tractors were real killers. They had a spring under them that pushed you up into the steering wheel. They're dangerous because of that. The only mark on Murray was a little broken skin on one of his fingers. We put his body on the wagon and hauled it out to Laidman Lake, where Floyd Vaughan was waiting with his floatplane.

SAGE: John still remembers the Vannoy kids. He met them at Rainbow Lake, when they were leaving the country with their mother after their father was killed. "The Ziglers had two boys my age, Brent and Jake," he says. "We were all in the swimming hole when one of the Vannoy boys took a crap in the water. We all boiled out of the swimming hole. It grossed everybody out and Dad was cussing and swearing. Those kids didn't know. They were hillbillies and had never been around people."

CORKY: That little house they lived in wasn't big enough to cuss a cat in, and the creek where they got their water was hot. It was warm water that ran down the creek from a swamp, and it had a lot of algae on top of it. That was their drinking water. Murray wasn't a hippy back-to-the lander: he was a religious nut.

JEANINE: We were at Rainbow Lake visiting John and Sandy when Denise Vannoy came out with the kids for the last time. They all stopped there. The Vannoys lived very isolated out there, but probably knew the Ziglers as well as they knew anybody. Somebody was with Denise, helping her drive the team. Corky and John Zigler had gone into the Blackwater with our tractor and wagon to haul any of Denise's things she couldn't take out with the horses. She was headed back to Montana where they were originally from. I remember thinking, "My God those kids are going to be so lost." I think the girls were born in the Blackwater. I'm not sure about the boys. The boys were seven and nine years old and the girls were three and five. The oldest boy might have already been born before they came there but he would have been very small. It was the only life they had ever known. I remember wondering how Denise was going to readjust. Her life with Murray Vannoy couldn't have been easy. She worked like a dog.

He was mental and his family bore the brunt of it. There were people out there living in the backcountry who couldn't accept the times they lived in. The whole reason they came there was to go back in time or to go to a place where they didn't have to deal with the world as it was. They wanted to make their own little world, and run it according to their own rules. Murray was one of those.

In a way we left a bad situation in Los Angeles and the United States too, but it's how far you went with it. We loved the isolation in one way, but we had a radio and books. Murray didn't believe in any books except the Bible. We kept up with the world, even if we weren't living in it and were fairly happy about that. We had a radio that was a big help to us for entertainment. Corky listened to the sports stations out of San Francisco and Oakland, and caught the football games. I loved the radio because during the day I could listen to CBC clear as a bell.

A SEASON OF FUNERALS

SAGE: Funerals in Aboriginal communities were significant events in the Chilcotin Plateau. It was not uncommon for a wake to last a week to ten days to give people travelling great distances time to get there. No matter what time of year, a death signified the call to gather. These potlatch affairs took precedence over everything else going on, whether it was haying or working for wages. People found ways to travel hundreds of miles at the drop of a hat, in all kinds of weather, to pay their respects to the departed and support the families suffering the loss.

In the summer of 1974 there were four deaths, one after the other, in the Native community, and four funeral potlatches. This created a dearth in the labour force that Corky was counting on to help harvest his hay crop. Eventually he didn't have much choice but to look outside the immediate area to get the help he needed.

CORKY: The summer of 1974 was a bad one for our Native crew. There were four funerals in a row and they had to leave and were gone for over a week each time. We had a lot of hay land to put up with the square baler, but since we couldn't get any of the local people to stay with the haying, we decided to go to Williams Lake. There was a youth hostel there, and we stopped by to see if we could get some hippies to work for us. The hostel had a labour board and we found some people eager to work. They jumped in my pickup so fast I couldn't even count them. I said, "I can only take three," and went, "You, you and you, get in. I'll feed you." They were hungry and we got some hamburgers.

One of the guys was Dana Langley, whose dad was a lawyer in Williams Lake. He later changed his name to Sunee Yuho and hung around the Chilcotin for several years after he finished haying for us. Dana would eat anything. He wasn't afraid to try things other people might find abhorrent and wouldn't consider putting in their mouth. Mosquitoes were one of his favourite delicacies and there was certainly no shortage of them at Muskeg.

Another guy we called Doctor Bush was a bit of a presumptuous character. Our neighbour Eustine Squinas was pregnant and about to have her baby and he wanted to go in there and help with the delivery. Fortunately we managed to talk him out of it before he upset them. The third guy went by the name of Tennessee. He didn't have any underwear and the seat was worn out of his pants. Fred would say, "See that guy with his ass hanging out?" Tennessee'd bend over and was the talk of the country. He was a big guy too—tall, easygoing, and he had gorgeous long hair, but he was missing a little upstairs. Not the sharpest nail in the box. He was very sweet-natured but might have done a few too many mushrooms. All three of them stayed the summer and worked hard, and we put up a big crop of hay.

For Jeanine, that whole summer was just one huge meal. We moved between Corkscrew Creek, Lessard Lake and Muskeg, and all of them could eat like horses. They worked hard and were a starved-out outfit when we first got them. Jeanine was making two or three pies at a time and they'd disappear. Fortunately, the price of food was reasonable compared to now. People didn't expect a whole lot. We gave them lots of good plain cooking, and Jeanine was an excellent cook.

Most of that summer Fred was off funeraling. You could get him out of Anahim Lake once in a while if you caught him in the right mood, but one of the people who died that summer was a member of Daisy's family, so Fred was mostly gone.

When we bought the Corkscrew place, D'Arcy had just bought a brand new square baler. There was some kind of farm program where the government would help buy you equipment, and he bought a baler because he was going to get away from putting up loose hay. He couldn't get the help putting it up, and that took even more labour and expertise than handling square bales. Stacking loose hay was quickly becoming a lost art. The summer of 1974 we put up a mountain of square bales with our hippy hay crew. The next year I brought the first round baler into the country, and that changed everything. We got away from having to depend on a big hay crew to get the square bales off the field before they absorbed water. The round baler

FRED ELKINS WITH HIS SON, GARREN, AND CORKY'S SON, JOHN, AT THE
OLD HOLTE CABIN AT MUSKEG.

revolutionized ranching. It got the hay up quick and you could leave it in the field until you got the rest of the hay in.

I also brought the first four-wheel-drive tractor and the first Brillion seeder into the country. The seeder had settings on it for a whole range of different kinds of seeds you might want to plant and a roller that packed the soil. Nobody had ever seeded those wild hay meadows before. I was the first one to rotovate them and break them up and seed them with tame grass. It grew like crazy once you burned the old hay off.

HAND-SCOOPING ROUND-MOUTH SUCKERS

CORKY: One time Fred and I went into Anahim Lake to get some supplies and Daisy fixed us some deer meat and Indian bread. We stopped at Christensen Creek along the way to eat lunch and I noticed that the creek was full of big springtime suckers. I asked Fred how his people caught those fish before they had all the modern inventions like fishing lines and hooks.

Without saying anything, Fred took off his shoes and jumped out into the creek up to his waist. He waded upstream and got a big rock and threw it out into the middle of the creek. He knew those big fish would go under the banks that hung down where the water had washed them away. Then Fred ran his hands along the bank and under the fish and slapped them right up onto the bank. He caught nine of them that way, and we brought them home to Daisy and she cooked them up. They were good-sized fish, and quite delicious.

A lot of people won't eat these round-mouth fish. But if you fillet them properly they don't have a bone in them. You throw the backbone out and you have a fillet of delicious white flesh. They are beautiful eating if they are cooked right.

Fred could only go so long before he had to go into town for a couple of days on a toot. He'd be working steady for a month and then he'd say, "That's the it." That was the expression he always used whenever something was finished, "That's *the* it," and that expression got loose in the country and everybody started using it.

I don't think the Carrier language distinguishes gender in pronouns like him and her. A lot of the old-time Native people used their pronouns interchangeably. When you first hear it, you wonder what they just said. Then you figure it out. One time we went out to cut firewood and Fred pointed to a nice stand of dead Jack pines on Crown land just off our property. "I don't think we can cut over there," I told him, "those are the Queen's trees." Fred looked back and said, "I don't think the Queen, he give a shit."

Fred and Daisy would pack up the kids when they wanted to go into town, and they'd be gone for a few days or a few weeks, depending what was going on. After a time they'd want to come back out and get away from all the partying. When they were ready to come home it sometimes took two or three days to get them all gathered. It was a ritual, and somewhat comical. Fred would track down Daisy and load her. Then she'd run off someplace, and Fred would take off looking for her and get lost himself. You had to herd them all back like a flock of sheep. I remember that well. That's just the way it was.

WHERE EVERYBODY LIVED

ANAHIM PEAK

SAGE: The broad floodplain estuary of the Dean River flowing north from Anahim Lake separates the multicoloured Rainbow Mountains to the west from the Itcha and Ilgatchuz ranges to the east. Standing over this landscape like a regal monarch is the basaltic plug of Anahim Peak. Corky and Jeanine's home at Muskeg Meadow sat at the foot of this landmark.

CORKY: The whole country around Anahim Peak is kind of spiritual. We could see the peak out our kitchen window at Muskeg. It was beautiful. It seemed like that mountain had its own unique weather system. Fog hung around all the time. Sometimes the mountain would be wearing a hat; other times she wore a dress. Every day it had a new look. "Anahim Peak got her dress on today," Big Fred would say. The Native people had gone in there and found obsidian behind Anahim Peak long ago. They called it Besbut'a, or Obsidian Hill. It was the only source of obsidian in southern British Columbia, and that volcanic glass was prized for making arrowheads, knives and spearheads. If it's prepared just right, it is the sharpest substance known to man.

There's a good view of Anahim Peak from Louie's place at Abuntlet Lake. It's beautiful standing there looking at the mountain and the river, the way that peak comes up like that.

NEIGHBOURS DOWNRIVER

SAGE: Three or four miles past Lessard Lake was Muskeg, where Corky and Jeanine established their main ranch headquarters. They hired Mac Squinas to build their new home there out of logs, and they

73

A VIEW OF ANAHIM PEAK ACROSS MEDICINE LAKE.

hired Andy Cahoose to build a second log cabin for Big Fred and Daisy. Andy and Annie Cahoose and Ken and Linda Karran lived nearby along the Dean River. They were the closest neighbours to Muskeg.

CORKY: Andy Cahoose and his family lived on the river side of the road past Lessard Lake. Ken and Linda Karran moved in right beside them soon after we came to the country. Muskeg was a couple of miles away on the side of the road away from the river. Below that was the old Clark place farther down along the river, which Lester Dorsey and Woody Woodward used as a hunting camp. Later, David Gladden and his two boys, Joe and Frank, moved in there for a couple of years.

DAVID GLADDEN

CORKY: David Gladden and his wife and four children had been living down in the Precipice Valley south of Anahim Lake before selling out to Lee Taylor in the mid-1970s. That's how Lee got in there. Gladden had a reputation for being quite eccentric, but he was all right. He was a lot smarter than people gave him credit for. He'd talk real slow, but he was a good guy. His woman was twenty years younger, maybe more, and she

DAVID GLADDEN AND HIS TWO BOYS, JOE AND FRANK.

ran off two or three times. Finally he let her and their two daughters go. He and his two boys ate out of tin cans. You'd go to his place at mealtime and somebody would open a can of peas and that was it. Nobody cooked.

I saw them in Williams Lake one time. I had to go see the doctor and David was in the lobby of the clinic with his two boys. He had all kinds of tinned stuff like corn, beans and carrots in his tote sack. He got hungry while they were waiting and he opened his sack, took out a can opener and a bunch of spoons, and opened the tin cans right there. He asked the people sitting there, "Do you want some corn? I've got peas if you don't like corn." The boys did it too. They got out their can openers and opened a can of peas. They'd eat a can of peas and then put it back in the tote sack.

ILDASH

CORKY: Three or four miles past Muskeg was Ildash, the last of the properties we bought from the Holte family. We found out later that cattle didn't seem to do too good down at Ildash. It had plenty of grass but there was no kick to it. Bob Cohen and I were down there one time glassing the mountain with binoculars from the meadow, and you could see coloured stuff way up high, some kind of mineral on the hillside that washed right down the mountain all the way to Ildash. So we got to wondering if the meadow was getting something from the mountain that was poisoning the grass. Tommy Holte had said something to Bob about that years before. A couple years later we sold Ildash to Gordon Richardson, and he later sold it to Peter Fuller.

SALMON RIVER

SAGE: Just past Ildash the road downriver forked. One branch headed east around the Ilgatchuz Mountains to Rainbow Lake and Irene Lake and into the Blackwater. The other fork continued down the Dean River to Salmon River where all the Cahooses lived. Old Joe Cahoose and his large family lived there. Joe and his wife, Mary Joe, had seven children. Their oldest daughter, Minnie, and her husband, Patrick Sill, had sixteen children and they occupied their own household there, as did many of their other children and their families. It's a big area with lots of meadows and places to trap.

All the members of Joe Cahoose's extended family hayed together in the big sprawling meadows that occupied both sides of the Dean River at Salmon River, where it wound back and forth like a snake. Susan Hance, the eldest child of Pat and Minnie Sill, says the family would put up twelve big haystacks using horses.

Beyond Salmon River the wagon road left the Dean River and continued north to Ulkatcho Village and into the Entiako country. At one time Ulkatcho families occupied the whole landscape all the way to the big lakes that form the Nechako Reservoir, utilizing the hay meadows, lakes and wetlands for their traplines, fishing and hunting areas.

CORKY: We didn't go down to Salmon River very much, but the Salmon River people always stopped by our place. We made an attempt to get along with our neighbours and fenced off a little bit of pasture for them right there at Muskeg, so whenever they came by and needed to spend the night, they had a place to put their horses. We had a watering place for the horses and they appreciated that. Once we moved into our new house we let them use the old Holte cabin any time they wanted.

THE CAHOOSE FAMILY BUNDLED INTO THE SLEIGH AT MUSKEG CABIN FOR A RIDE DOWN TO SALMON RIVER.

SANDY ZIGLER AND THE ACCIDENTAL GOAT-KILLING

JEANINE: We often visited John and Sandy Zigler because they were our nearest neighbours along the road into the Blackwater. Their two boys and our son were close in age, and we enjoyed socializing with them.

John Zigler had decided they were going to have a farm there, and he had a growing collection of farm animals that included two wonderful pigs named Rhoda and Tiller. They really did rototill that place, and they were helped out by a bunch of little piglets running around. The Ziglers also had goats.

In Diana Phillips's book *Beyond the Home Ranch*, she mentions that Sandy Zigler shot at their goats, and I remember it well because I was there.

We were visiting John and Sandy at Rainbow Lake for a couple of days, and John had just acquired a couple of nannies, which were quite nice, but he also had the nastiest, smelliest, meanest old billy goat that you ever did meet. The creature was horrible. It's no wonder that when they paint pictures of Satan they give him goat eyes. The billy loved to sneak up and hit people from behind, and if he caught you in the outhouse he would run at the building and scare the living daylights out of you. We all hated this goat, but John was determined he was going to raise goats.

SUSAN HANCE LEARNS THE OLD WAYS

SAGE: While a disconnect from traditional skills and values was the trend, the family of Pat and Minnie Sill saw the need to ensure that knowledge of the old way of life wasn't lost. They made sure that one of their offspring learned the old ways. Their oldest child, Susan Hance, was born in 1952 and grew up in Salmon River, downriver from Corky and Jeanine's place at Muskeg. She was the oldest grandchild of Joe Cahoose, who Corky so much admired.

SUSAN HANCE: I grew up mostly at Salmon River and Uskisula, which was my uncle Frank Sill's place, straight back behind Ulkatcho Village. My grandpa Thomas Sill's place was at Kiston, way down in that country. I've got a map showing all the old traplines.

Things happen to me in strange ways. Almost like the great spirit pressures me. I don't know. There was a "residential" dormitory in Anahim Lake for the Ulkatcho kids going to school. A whole bunch of us stayed there and went to school right up to grade seven. All the kids that finished grade seven were sent to Williams Lake for high school and they all stayed at St. Joseph's Mission. The truck would come along in September to pick the kids up and take them to the Mission. Dad called it a cattle truck. They were loading all the kids who finished grade seven into the truck and we were all getting ready to go to town, then out of nowhere Mom showed up with the team and wagon. "You're not going," she told me. "You will probably learn more if you come home with us."

At the time I felt really robbed of my education. I had no choice and had to go back home with my parents to Salmon River. Now I'm really glad I did that. I never made it to residential school at Saint Joseph's Mission where the kids from Anahim Lake stayed. When I went home my parents and grandparents taught me how to trap and hunt and speak the Carrier language and tell the old stories.

MINNIE AND PAT SILL IN A HORSE-DRAWN WAGON.

A few years later I met my husband, Ray Hance, down in Armstrong. In the mid-1960s this guy had come out to Anahim Lake to recruit First Nations people to leave the reserve and go picking and working in the orchards down in the Okanagan. Our family was one of them. I went down there and that's where I met Ray. We were picking asparagus, lying on a platform like airplane wings attached to the front of a tractor. I was only eighteen years old when Ray and I got married. When we first met I told him how I felt robbed of my education. He laughed and said, "You're lucky. You better thank your parents." Now that I know what's going on I feel so lucky my parents insisted I come home with them.

Our daughter, Carmen, was born in 1972, and our sons, Jason and Ryan, came along a few years after. Raising my kids, my belief was you couldn't leave them with a babysitter. So I stayed home with the kids to raise them. But as soon as my youngest one graduated from high school, I went after my schooling again. I went to Thompson Rivers University and took extra courses and finished grade twelve. I wanted to be a loan officer in a bank when I finished grade twelve, so I got a job as a teller at the CIBC. People from my community would come into the bank. They were so proud somebody in the bank could speak their language. When one of the other tellers asked if she could help them, they always said no, they wanted me to help them. A position for loan officer opened up and I applied for it.

At the end of August my sister Gloria Izatt and Corinne Cahoose showed up and said they needed a language teacher over at Anne Stevenson Secondary School. They said I had the language skills and would be a good person for the job. I could speak the Carrier language but wasn't sure I could teach it. Besides that I wasn't too interested. I liked working in the bank, and had my hopes set on getting the loan officer position, so I flatly refused them. Gloria and Corinne had to cancel the Carrier language course that semester because there was no one to teach it.

Toward Christmas I lost my chance to be a loan officer at the bank, and I was really upset because I really wanted the position. Then Gloria and Corinne came to see me again and said they really needed a high school language teacher. Without even thinking about it I said, "Okay, I'll do it." When I went to teach my first class, there were twenty kids in there. I thought, "Oh my God, what have I got myself into!" But it didn't take me long to get busy. I ended up teaching at the elementary school and high school. Besides teaching the language, I shared the stories and history of our people. I just love it.

John and Corky had gone off to do something, and Sandy and I were doing the laundry. We had a ton of it, as usual. We finally finished up the first round of clothes and hung them up, and then we sat down to a cup of tea. All of a sudden Sandy looked out the window and let out a shriek. The first thing I thought was that it was a bear. She jumped up and started cussing, "That blankety-blank goat!" I looked out and the old billy was tearing down the laundry, spreading it around the yard and stamping on it.

John had thought of a plan to keep the goat at bay if it kept doing things like trying to knock the outhouse over. He had loaded up some shotgun shells with rock salt, and the blast would inflict a little pain to get the attention of the animal and scare him away but wouldn't really hurt him. Sandy jumped up and ran into the other room, grabbed the shotgun and grabbed what she thought were the rock salt shells. Then she ran outside, loaded the shells into the shotgun and blasted the goat from about twelve feet away. He dropped like a rock without a struggle. He was a big goat and he fell over. The way he went down, we were hoping maybe he was stunned. But he wasn't: he was dead. He fell over and died with his feet sticking straight up. Obviously Sandy had picked up the wrong shotgun shells by mistake. There she was, standing over the goat saying, "Oh my God, I've killed the goat. Johnny's going to kill me, I've killed his goat!"

There was nothing we could do. We couldn't hide the goat, it was there in the front yard. When John and Corky came back, John wasn't mad at all. There was no big scene or anything. But oh, how we laughed about it. I think secretly John might have been wishing that the goat would die or get killed so he wouldn't lose face on the goat idea.

When John and Sandy left the country a year or so later, they gave us two turkeys, four geese, a hog and a goat. We had no place to put any of these animals because we were still getting settled at the Muskeg place, but we hated to say no. We were neighbours, after all, even though we had no pens for these creatures. The coyotes ended up killing one turkey and all four geese. One day one of the turkeys attacked Dana, who was just two years old and toddling around the yard. The turkey flew up and came down on her head. Luckily she was wearing a really thick wool toque. He had his claws right into her head and was whacking her with his wings. I'd never seen a turkey do anything like that before. I shrieked and grabbed a broom and whacked him a few times on the head and he got off and left. The geese would run at you too, but with them it was mostly show. They are great watchdogs. But that turkey, I was not sad to see the coyote make a meal of him. I didn't care.

As it turns out, we never did eat any of those animals. We butchered the pig but the weather was still too warm and we couldn't hold the meat and it started to go bad. One day Ron Gauthier and his crew showed up.

We thought the meat was a little off, and despite our warnings they ate it anyway and got sick. So we threw the rest of the pork out.

We gave the goat and the remaining turkey to somebody, after the geese and mean turkey were killed by the coyotes within two weeks. There was nothing left but little bits of feathers.

WINTER AT LESSARD LAKE

JEANINE: In late fall 1974 we moved up to Lessard Lake to spend the winter. Dana was just over a year old and still in her playpen when we moved in, and it was cold in that house. The two-storey house was built out on a point in the lake to keep the bugs down, but that also meant it was exposed to the winter winds. When you were lying in bed you could look up at night and see the stars shining through the cracks in the ceiling. The windows were so loose that when the wind blew, the curtains would stand straight out. Without the big barrel stove we couldn't have survived in there. The heat from the stove would melt the snow on the roof and the water would run off and in underneath the eaves and down the logs. Then the water would freeze on the inside of the wall and create a frozen waterfall two floors high. When you'd get a good fire going during the day to get the place warm, the fourteen-foot icicle would start to melt and water would run inside the house. There was always a little puddle below it on the floor.

Poor Dana had to wear so many layers of clothes. She always wore a little hat and socks in the house. I was so happy to get out of there. Even the old cabin at Muskeg was better than that. It was so much smaller you could actually heat it in the winter to where it was comfortable.

We had Lester's team of horses at Lessard Lake, and fed them loose hay that Mac Squinas had cut. I didn't like to leave Dana alone in the house when I went down to feed the horses, so I would dress her up in this little pink, fuzzy snowsuit that made her look like a little pink blimp, and I would take her with me. I'd make a little place for her up in the haystack while I fed the horses. I remember beautiful winter days with the sun shining and the blue sky, and this little pink baby waving her arms, stuck up there on the side of the haystack. It was one of those moments when you wouldn't want to be anywhere else. Then in late February we took the team and sled on the winter road and spent the rest of the winter at Muskeg.

CORKY: Knowing how to drive a team of horses was pretty helpful on a ranch. When it was forty below and the tractor wouldn't start, you could always harness up the team and they would go every time. Learning the ropes to work with the horses took a bit of getting used to for me. I had a wreck the first time I tried driving them on my own from Lessard Lake to Muskeg. Fred had come down early that morning and we got the sleigh loaded with everything we needed, and

BIG FRED ELKINS, WITH JEANINE AND THE KIDS, HOOKS UP THE TEAM AT LESSARD LAKE.

Fred hooked the team up for me because I didn't know how to do it. Then he got on the tractor and took the longer, windier road to Muskeg, a distance of about six or seven miles. I was going to take the horses down the shortcut that Lester had shown me over the frozen lakes and swamps, which was about half the distance.

Lester always said before you head out with the team, let them make a circle and run around a little bit so they get used to you handling them. Talk to them and let them get familiar with the load. So I did that. Then I turned them and headed down the winter road to Muskeg. We had to go right by Mac Squinas's place and he had some wild, half-starved dogs hanging around there. When they heard the jingle of the horse harness, they figured they might get something to eat, so they ran up and scared the hell out of the team. They were snapping at them, and the horses didn't like that worth a damn, so they kicked at the dogs barking and nipping at their heels. Then one of the horses got out of time with the other one and slipped and fell, and the sleigh lurched to a stop. The other horse panicked and fell in there too. In all the commotion the load shifted and I got hit right in the back with the washing machine. I'll never forget that one. It was the damndest mess you've ever seen with these dogs raising old Billy Hell.

Finally I got things settled down and I had a look around to see what was alive. I got off the sleigh and started talking to the horses. I untangled them from the mess and got them to their feet and once they had calmed down, I backed them away from the wreck. The main thing I was concerned about was the horses. They were Lester's pets and I didn't want them hurt. He always had me winter his team because I fed them real good and he appreciated that. When I told him about what had happened he said, "Man, you're lucky to be alive, but you must have done the right thing." I said that was the only thing I could do. When you have a bad experience like that, a lot of times it's for the good. It was all part of my steep learning curve moving farther into the backcountry.

I was completely green when it came to hitching up the team. When I didn't show up at Muskeg, Big Fred came back and showed me how to adjust the length of the tugs so that everything stayed together. With Fred's help I eventually learned how to work with the horses. Using workhorses was one of the romantic ideals that was so much a part of that country.

BIG FRED HELPS CORKY SKIN THE MOOSE

CORKY: One time we didn't have anything for Big Fred to do at Muskeg, so he headed into Anahim Lake. There was snow on the ground and I was checking the fence in the little haystack meadow at Gloomy Point when I surprised a moose eating willows. I had my rifle with me and had a pretty good shot. All of a sudden I had a thousand pounds of meat to deal with and decided I could use Fred's help. We sent a message to D'Arcy Christensen to pick Fred up and fly him down to Muskeg. I had the moose mostly gutted by the time D'Arcy landed with Fred in his ski plane. Fred and John took a chain and the tractor and dragged the whole thing to the house.

We butchered it up on the other side of the little pond so we didn't leave any blood, hide or meat scraps around to call bears down to the house in the spring. I really appreciated Fred's expertise when it came to handling a moose. He was so strong he could lift a whole moose quarter at a time.

I also needed Fred's help to cover up the evidence of butchering the moose on the snow. It was shot out of season and we didn't want to attract the attention of the game department or RCMP if somebody spotted it from the air.

JOHN WATCHES AS BIG FRED FINISHES SKINNING THE MOOSE.

JEANINE: When we first moved to Anahim Lake, Fred shot a moose because we needed meat. We got this moose out of season and I'm sure we didn't have a hunting permit. Corky was taking advice from the locals, who were saying, "Hey, let's go out and shoot a moose." That's something everybody did back then. There were lots of moose around and if you needed some meat you just went ahead and shot one. We brought it back, butchered it and hung it in the old barn there. There wasn't a regular game warden in those days. The game department would contract with a local person to do the game warden duties, so things were pretty relaxed when it came to hunting meat for your table.

The guys were out there still chopping the moose up and there came the truck with the game warden seal on it. It was hilarious. Everybody scattered like a covey of quail—Big Fred had the heart and liver in his arms and was getting ready to take it to his camp.

CORKY: Somebody hollered, "It's the bloody cops," and there was the cop coming up the driveway. I saw Bob Cohen sprint into the outhouse in record time, and Fred flung the heart and liver he was

CORKY AND ANDREAS BEHM SURVEY THE FINE HAY CROP AT MUSKEG.

carrying and they ended up somewhere on the other side of the shed. He threw it right over the damn shed roof, clear over the saddle house and harness house, out of sight.

When I saw the cop coming I went up to him and told him we were butchering a cow we had hanging there for our winter meat. He wasn't too worried about it. He said, "If anything happens around here you need me for, just call me up." He was the best cop we ever had.

THE SUMMER OF 1975

SAGE: By early spring of 1975, Corky and Jeanine had finally settled at Muskeg for good. After surviving the wreck with the horses, Corky took two decisive steps that greatly influenced their future. First, he commissioned Mac Squinas to build a new house for them at Muskeg, and Mac started that summer by cutting house logs and peeling them so they could dry sufficiently for construction the following summer.

Second, Corky invested in farm machinery that would transform the way he was doing things on the land. He traded in the relatively new square baler that had come with the Corkscrew Creek Ranch and bought a round baler. He also acquired a rotovator and Brillion seeder so he could start farming the wild grass meadows and grow more productive hay crops.

"It was the summer of the hippy hay crew that convinced Corky to go for the round baler," Jeanine says. "Corky wanted to develop more hay land, and the whole business of farming the swamp was trial and error. Corky was going into serious farming of the land."

The transition from operating the square bale equipment to running a round baler took some getting used to. For one thing, if something went wrong, they were a full day's travel away from the farm equipment supplier in Williams Lake.

"It took Corky a while to learn to use the round baler," Jeanine says. "Angry, frustrated, screaming obscenities. 'Bring me the whisky!' Those early balers had some problems with the strings that took some figuring out. I'd look out the window and see the baler stopped. I'd think, 'Oh boy...' Then I'd bake pies. Corky would get so mad."

FRED AND DAISY MOURN THE DEATH OF THEIR BABY

CORKY: Big Fred and Daisy lost a newborn around the time Jeanine and I first came to Anahim Lake. I don't know what happened, but shortly after the baby was born, somebody gave Fred and Daisy and their baby a ride to the hospital in Bella Coola. There's a huge rock on the side of the road in the Bella Coola Valley about ten miles from the bottom of the Hill, and there's a pullout there. The baby died right there in that pickup as they were heading down the road.

JEANINE WITH HER DAD, JOHN SEALS,
HOLDING LITTLE JOHN.

The year we got the round baler we got through haying much earlier than we ever had before. Fred and Daisy wanted to go fishing in Bella Coola, so I drove them down there. When we got close to that big rock, both of them started crying and they wanted me to pull over. I did, not realizing what was going on. I was just taking them down there. When they saw the rock, it made them think of their baby. Once they told me the story, I couldn't help but cry a bit myself.

SENDING JOHN TO SCHOOL IN TEXAS

SAGE: Jeanine started homeschooling John, then in grade one, during the winter they stayed at Lessard Lake. At the end of the summer in 1975, her dad, John Seals, flew up to visit, and he took Little John back to Texas with him for grade two.

"John was named after my dad," Jeanine says. "He went to stay with Corky's mother, Vivian, in Crosbyton, in the high plains of the Texas Panhandle near Lubbock. I missed John terribly. Later that winter we all drove down to see him. Dana and I spent a couple of months at Vivian's with John, while Corky returned home after a couple of weeks to look after the ranch."

Corky's parents, Jim and Vivian, were in the process of splitting up when Little John went to stay with Vivian. "Corky's dad had already moved out when my dad flew John down with him to go to school that first year," Jeanine says. "Corky was upset over their separation and never got along with Jim's new wife, Adele. They came up to visit us at Muskeg once we built our new house and Corky had a definite bias against her. He made up his mind he wasn't going to like her because she had busted up his parents' marriage."

VIVIAN, DANA AND JEANINE IN CROSBYTON.

Jeanine says having John stay with her so soon after her marriage to Jim ended made a big difference to Vivian. "Vivian just loved having John," Jeanine says. "She just loved him so much. She would go to church on Wednesday night for prayer meeting and John would stay in the house by himself because the church was just down the road from the house. John would call her at the church and tell her that he was going to bed now, and she'd tell him that's good and she'd be home in half an hour. John really enjoyed it down there too. He had lots of little friends he played with. Vivian knew everybody in town since she was a schoolteacher, and she made sure he'd have friends over."

John went to school in Texas for two years, staying with his grandmother both years, but it was torture for Jeanine and Corky being separated from him. The second winter, Jeanine and Dana went down again and spent a couple of months with John at Vivian's. "It worked out, but I missed him bitterly," Jeanine says. "Corky and I both missed him a lot."

Two years of sending John out to Texas for school was enough and Jeanine did another year of homeschooling for grade four. "The following year we took him to the dorm in Anahim Lake where the Native

CORKY AND JOHN ON OL' YELLER; DANA STANDS BEHIND THEM.

kids from downriver were staying, but he hated it. He spent a month there, then we took him home and homeschooled him again."

Eventually, in the fall of 1979, Jeanine made the tough decision to move to Williams Lake so the kids could go to school. She found a place to live at Rose Lake, halfway between Williams Lake and Horsefly, and both kids enrolled at 150 Mile School.

BURNING THE BILL LEHMAN MEADOW

CORKY: As part of the land exchange with the Holte family, we acquired an old yellow 1940s-vintage Massey Ferguson tractor that I named Ol' Yeller. Ol' Yeller wasn't much to look at, but she was hell to go. The Holte boys had stripped her down for speed. The hood was gone, as were the fenders, and the starter was burned out. You had to pull it to start it. Big Fred called it the tractor that would not die. This tractor was great in mud and had lots of clearance, so you could take it where other tractors couldn't go.

Bill Lehman Meadow grew hay six feet tall, but it hadn't been hayed for years. It was wet most of the year, except in the fall. Millions of mosquitoes lived in the old grass. When Fred and I walked through the old grass, swarms of these vicious biting sons of bitches came out just like smoke. In the fall of '75 we decided that fire was the only solution. Burning the dead grass would remove the bug habitat and add lime and potash to the soil. So we did the only logical thing: we got drunk and set it on fire.

I bought a rotovator for Ol' Yeller and me and Big Fred rotovated a fireguard around the outside edge of the Bill Lehman Meadow before setting the grass on fire. After we burned off the meadow, we could see how good the land was. The meadow gently sloped toward a creek down the middle. With the grass burned off we could see the high and low spots that could be levelled by rotovating and dragging a spike-tooth harrow over it. We laid out a nice piece of ground, approximately fifty acres, and started rotovating. Then we harrowed it. We had to work fast because it was late October—we knew the ground would soon be frozen and we'd be forced to quit farming. On the last day of October we went home to Muskeg to pick up some equipment and see our families, and that's when the Big Snow hit. We didn't know it then, but we were going to be right in the midst of one of the worst snowfalls in the history of that country.

THE STORM OF '75

SAGE: The snowstorm of 1975 was one of those events in the West Chilcotin that people still talk about and use to measure off against time. Over a two- or three-day period at the end of October into early November, more than four feet of snow accumulated on the ground. Strong winds created massive snowdrifts and a mishmash of fallen trees across the landscape blocked roads and trails. It wreaked havoc and caught everyone by surprise. Guide outfitters off in the mountains with their horses and hunters were stranded. Ranchers still had livestock scattered across the range, and horses that normally rustled freely all winter on wild forage were suddenly immobilized and cut off from access to feed. There were heavy losses of domestic livestock and wildlife that winter.

When Corky and Big Fred left Ol' Yeller in the field at Bill Lehman Meadow, they were planning to resume farming the next day. By the time the snow quit falling, the old tractor was completely buried from view. Ol' Yeller remained where Corky and Fred had parked her for the next year and a half, bogged in the quagmire.

When Fred and Corky headed back to Muskeg, Corky was expecting guests from Williams Lake later that day. His banker was coming out to go moose hunting.

JEANINE: The winter of 1975 was the worst of any that I can remember. It wasn't the cold, because it was actually very mild. It was the snow. The first inkling we had of what was to come happened at the end of October.

We did our banking at the Bank of Montreal and our loan officer was a very nice guy named Marvin. That fall he asked Corky if he and a friend could come out one weekend and go hunting. It's always good to keep your banker happy, so of course we said yes. Corky explained that it was pretty rough out there, and that we were living in a fifty-year-old cabin at Muskeg Meadow that had been built by one of the pioneering families in the area. It was basically one room and the roof was made the old way, with split poles and shakes and a layer of sod in between, and it leaked like a sieve when it rained.

Marvin assured us that he was a country boy himself and would not be put off by an outhouse and water packed from the creek by bucket, so we told him to come on out. When he and his friend arrived we had a little snow on the ground but he had a big four-wheel-drive truck and had no trouble getting into the place.

That evening it began to snow big, wet flakes. We didn't think too much of it and hoped it wouldn't make it too hard for the guys to get around the next day. During the night the wind got up and we began to worry a little. It was blowing from the west, which was unusual for Muskeg. When we got up the next morning we really started to worry. The snow had not let up at all and it was over a foot deep. After we talked it over, Marvin decided that they had better head back to town as soon as the snow eased up enough to see the road. Corky went to get the tractor so he could break trail down to the road. Unfortunately, the snow didn't let up; it just got thicker.

By this time we knew we were in trouble. It was even getting hard for the four-wheel-drive tractors to get around and we had to start feeding the cows. Marvin was starting to pace the floor, though there wasn't much floor to pace. He realized there was no way for them to get out unless the road was ploughed. Unfortunately, there was no such thing as regular snowploughing on the Dean River road. Normally we

THE SNOW REACHES SHOULDER HEIGHT DURING THE STORM OF 1975.

bulled our way through as long as the rig would make it to town. After that it was horse or snow machine. Some people would simply leave their rigs in Anahim Lake for the winter. We knew that the folks in Anahim would realize the situation and get us some help, but we didn't know how long it would take.

The snow kept falling. The temperature was barely below freezing and this made the snow wet and heavy. Despite the difficulty, we knew we had to get up on the roof and scrape the snow off before the roof collapsed. About that time we started hearing a sound that would become very familiar over the next few days. It started with a creaking sound and ended with a huge thump you could almost feel. The trees were starting to fall.

Loaded with heavy snow, they could not resist the force of the wind and they were being torn up at the roots. Others didn't go down completely, but lost large branches or snapped in two halfway up the trunk. This blowdown would make every trail and road a hazard for travel.

It was a serious situation but not without its funny moments. John and Sandy Zigler had pulled out of the country that fall and given us a hog, which we had planned to butcher as soon as the weather got cold enough. Because we weren't planning to keep him over winter we had not built any shelter for him and he was miserable in the snow. The first night he tried to solve his problem by digging under the cabin. When he first attacked this project we had no idea what was going on. It sounded like a bear was trying to tear the back wall out.

Corky got his gun and we all moved as far as we could to the other end of the cabin. It was then, while I was listening with every nerve alert, that I heard something familiar about those sounds. About the same time Corky looked over at me and burst out laughing. "It's the pig!"

Everybody settled down once we figured out what was going on. I felt sorry for the hog but at that point there wasn't a thing we could do for him. He kept it up, on and off, for the rest of the night and it certainly didn't make for a peaceful rest. The next day we shovelled out a path to Fred's cabin where the porch was high enough that he could burrow under. We fed him and put some hay under there for bedding, and all was well.

The pig wasn't the only creature to cause trouble. The roof of the cabin was prone to leaking anyway, and with all that wet snow it was dripping everywhere. I put down all the pots and pans I could spare, but that didn't work too well as I had to keep taking them back to cook with. I finally got out a roll of heavy gauge plastic and cut off pieces to cover the food supplies so they would not be ruined. This plastic was fairly thick and it rustled and crackled whenever it was moved.

We went to bed early that night as nobody felt like talking, and besides, we didn't know how long our lamp oil would last. During the night I woke and lay listening to the leaks going plink, plink, plink. As I lay there I got to worrying about one section of plastic that hadn't seemed securely fastened. Finally I realized that I was never going to get back to sleep until I checked on it. Not wanting to wake anyone, I didn't use a light and tried to be as quiet as possible. Unfortunately, that wasn't very quiet because of the plastic rustling every time I touched it. Satisfied that the plastic was secure, I tiptoed back to bed.

I had just snuggled in and realized that Corky was awake too. Then Marvin called out in a rather concerned voice, "Corky, I think there's an animal in here."

"What's that?" Corky asked.

"I couldn't see very much, but it seemed pretty big and was sort of fuzzy," Marvin quavered.

"Well, I think it was just a packrat, but I'll light the lamp if you want." After a small hesitation Marvin said, "No, it's all right."

I'm sure the fact that Corky is a trained actor let him say that last line so calmly. I had my pillow crammed in my mouth to keep from howling with laughter. It just so happens that I had very curly hair and when it was humid it looked like a giant fuzzball. Big and fuzzy!

By morning a real sense of desperation was setting in. Marvin and his friend knew that their families would be worried sick by now and there was no way to let them know they were safe. The snow had finally stopped falling but it was so deep that moving anywhere was almost impossible. We had dug out paths to the creek, chicken house and outhouse, and the snow was as high as my shoulders. The cabin was almost buried and we had to keep the door shovelled out so we could get in and out. We also had to clear the windows because the snow was so high that you could literally walk over the cabin. The wind finally died down and the air was so clear that sound travelled a long way.

At first, I wasn't sure I was hearing something. I was afraid I might be imagining it because I wanted to so much. But it was a real sound, some kind of big machine over on the road. I yelled and everybody came scrambling out of the house like prisoners let out of jail. It was an hour before we heard it turn onto our road, and finally we saw the most beautiful sight we could ever imagine. It was George Reed on his Cat, ploughing us out. It turned out George had been down at Salmon River doing some work for the Indian Department and got caught in the blizzard, too. He was on his way out to Anahim Lake and was ploughing out the places along the road as he came. He had been on the road for eight hours and had a little caravan of people who had been snowed in waiting on the road for him, including several guys who had been hunting down the river. Marvin practically spun out in his boots, he was moving so fast. They had their stuff packed and were ready to go before George finished his cup of coffee. It took another ten hours for George to deliver them all safely to Anahim Lake.

The fact that we had a toddler in the midst of the big snow, getting into everything as only a two-year-old can, added to the claustrophobia we felt in the small cabin. I remember taking Dana out—the paths that we had shovelled were way over her head. She would be trundling along and would fall into the snow, unable to get out without help.

SAGE: Jim Schurr lived at Towdystan with Fred Engebretson, fifteen miles south of Anahim Lake along the Chilcotin Highway. He says the Big Snow impacted that part of the country as well, pushing all the buckbrush right to the ground. "We had a hell of a time ploughing it at first because the snow was too sticky and wet," he says. "That kind of snow that sticks to everything is hard to handle. In this country the snow is usually pretty dry and powdery. Once it gets cold enough, the snow turns to sugar and runs off the blade. After that deep snow it wasn't too bad a winter, but at Towdystan we didn't manage to put up very good hay, and we lost quite a few calves that spring. Fred [Engebretson] cut a bunch of poplar trees and the cows ate the bark, but I think it aborted some of the calves. It was a pretty tough winter."

CORKY: After the Big Snow, the temperature actually warmed up that winter. The hay under that snow stayed just as green as a gourd all winter. I expected it to be like straw, but it stayed green. Most of the winter nothing could get to the grass to eat it because the snow was so deep, which kept the ground from freezing. By the time it stopped snowing you couldn't see the hay bales any more. We knew we had lots of big round hay bales out in the field, but you couldn't see them.

After the record snowfall we were looking forward to spring. When spring finally did come, it came with a vengeance. All that snow melting off the Ilgatchuz Mountains had to go somewhere, and a lot of it ran right through Muskeg on its way to the Dean River. The country from Cless Pocket to Salmon River was underwater. It was one big lake. The Native people said they had never seen anything like it in their lifetime. Fences and stackyards were underwater, and many round bales floated off, never to be seen again.

DIGGING OUT AFTER THE BIG SNOW

CORKY: When we began to dig out from one of the worst snowstorms in BC history, we didn't even know where to start. A lot of trees had fallen across the road before George Reed made his pass with his bulldozer, and a lot more fell across the road after he had gone through. The whole landscape was a broken Jack pine jungle that was impassable unless it were ploughed or cut out.

It snowed up to the eaves of the house every night and Big Fred and I had to shovel the snow off both my cabin and his cabin every few hours to keep the roofs from caving in. Once we got the snow off my cabin, we covered the roof with heavy-duty plastic tarps, and threw some old tires on top to hold the tarps in place. Of course to get the tarps we had to shovel out the tool shed, which was completely covered with snow. Eventually we cleared the snow from around all the out-buildings, only to have them buried again the next day. Then we had to go out and find them and clear them off again. One of the first priorities was to shovel out the outhouse, but first we had to find it. Fred cut a Jack pine pole about fifteen feet long to probe with. Once we found the outhouse it had seven or eight feet of snow drifted on top of it and took several hours to dig it out.

Next we had to find all our machinery and dig it out. Fred took his Jack pine pole and started to probe for it as well. Our mowers, hay rakes, round baler, a couple of pickups, two wagons and a variety of other farm machinery were under at least ten feet of snow where it had drifted. The haystacks were the most serious because the snow had to be removed from them or the hay would rot. We had several stacks that had to be cleared off and it was a lot of work, but we saved our hay and found our machinery.

Ranchers and people living downriver weren't the only ones affected by the Big Snow. It was hunting season and several guides and hunters who were snowbound in the backcountry had to be rescued by helicopter.

With the deep snow and deadfall impeding access, the whole country was paralysed. The snow was too deep for the horses to paw through like they normally did in winter, and they became exhausted and literally lay down and died. The storm came so suddenly and so early that many cattle caught on the range couldn't get back to their home ranches where ranchers had hay to feed them. They also gave up and their bones are there yet.

After the snow settled a couple of feet, Big Fred and I decided to dig out our snow machines and check on Ol' Yeller to see if we could get it out. We couldn't even find it, so we left it there all winter.

THE BIG SNOW

WINTER ROAD

CORKY: The winter road from Muskeg to Anahim Lake was much shorter than the summer road because we could take shortcuts over the lakes and through the swamps once everything was frozen over. One person would break the trail and everyone else would follow.

After a certain point, when the snow got too deep, you parked your vehicle for the winter because you couldn't get out. A lot of people would park their rigs in Anahim Lake and travel back and forth by snow machine. Our Toyota Land Cruiser was the best pickup I have ever driven. It was real good in mud and snow and you had to get bogged down awfully deep before you got stuck. Dogan Leon was quite impressed with her. I can still hear him with his high-pitched voice and Ulkatcho accent: "Cogie, he go! Wintertime, he go! Cogie go!"

MUD BOGS

CORKY: The Bubble Gum Hole at Cless Pocket Ranch where Christensen Creek runs through was pretty bad. That's where Fred and I were having lunch when Fred scooped those fish out of the creek with his bare hands. Before Bryce and Sherry Sager bought Cless Pocket, the road went right by the house. Bryce got tired of pulling people out of the Bubble Gum Hole so he moved the road away from the creek. It was called the Bubble Gum Hole because the mud and clay were sticky and elastic just like bubble gum. There was another bad mudhole about four or five miles farther down the road we called the Moccasin Bog Hole.

Our neighbour Bernie "Burnt Biscuit" Wiersbitzky used to travel through our place at Muskeg all the time with his horse and wagon to his ranch way off downriver in the Entiako country.

L: JEANINE, LITTLE JOHN AND BUTTERBALL TRAVEL BY SNOW MACHINE.
R: TRUCK CHAINED UP FOR WINTER TRAVEL.

BERNIE'S EPIC STORY OF SURVIVAL IN THE BIG SNOW

SAGE: Bernie Wiersbitzky lived four or five days' journey downriver by team and wagon past Corky and Jeanine's place at Muskeg. He got the nickname "Burnt Biscuit" because he was a cook by trade and his first job in the country was running Ivy's Cafe in Nimpo Lake. Everybody in Anahim Lake country had a nickname, and Burnt Biscuit was a fitting appellative—a lot easier to say than Wiersbitzky, and much easier to spell.

Bernie often stopped by to visit Corky and Jeanine and rest up his horses while coming or going between his ranch and Anahim Lake, more than one hundred miles to the south. His place was beyond the Entiako River and close to the Nechako Reservoir.

In 1975 there were a number of Ulkatcho families still living along the route Bernie travelled, running their traplines and feeding their small herds of cattle on isolated meadows over the winter. During the year of the Big Snow, Bernie had his own perilous adventure and lived to tell about it.

He left Anahim Lake with his team and wagon in the pouring rain during the last week of October, heading for his ranch. The wagon road past Cless Pocket Ranch was basically a bush trail, where he couldn't do more than twenty to twenty-five miles a day, and by the time he got to Gene Jack's place, the last occupied cabin before his ranch, there was a foot and a half of snow.

BERNIE: I had to abandon the wagon at Halfway Meadow, which is halfway to the ranch from Cold Camp Lake, because the snow was so bad. The snow was three-quarters up the wheels of the wagon and the horses were played out. It was dark and I thought to hell with it, I'm going to unhitch the horses and get home tonight. It gets dark around five o'clock at that time of year, and I got home about eight, walking the horses that last five or six miles to the ranch. In the following days I made a number of journal entries documenting my experiences in the Big Snow.

> **SATURDAY, NOVEMBER 1**—This evening I arrived at the ranch without the wagon. I had to unhitch on the other side of Halfway Meadow because there was quite a bit of snow and very hard pulling.

> **MONDAY, NOVEMBER 3**—Today it snowed real hard, about one and a half feet of wet snow. In the morning it was deathly quiet. I looked out the window and the hitching post in the front yard was starting to get buried in the snow. This much snow meant I couldn't get the wagon this year.

> **WEDNESDAY, NOVEMBER 5**—Today I broke trail to the wagon with the team. It took all day to get there. I spent all night in the wagon without a fire and I was wet. The snow was falling all night. The return trip along the broken trail only took three and a half hours.

I had to bring my horses out to Anahim Lake through the Big Snow because my hay meadows were so far underwater that summer that I wasn't able to cut much hay. I had a little haystack, but not enough for eleven horses. I left on the tenth of November and I wasn't at Corky's place until the eighteenth.

> **MONDAY, NOVEMBER 10**—Today I am leaving with all the horses to go up to Corky Williams's place. The weather is turning cold. It's zero Fahrenheit. I was only able to go as far as Entiako River and spent the night there.

BERNIE FORDS A RIVER WITH HIS TEAM.

What usually took me three days, packing horses from the ranch to Anahim Lake, took me nine days. I only had food for three days. I spent two nights at Salmon River and then I came to Corky's place.

With great difficulty I had to lead the horses most of the way. I got as far as Cold Camp the first night and spent the night at the Entiako River. The next night I got to Gene Jack's. Gene had just come in and he said the snow wouldn't be so bad once I got to John and Maddie Jack's place at Majuba, eight miles away. He figured Maddie and John would have come in by that time with their sleigh. "You'll have easy going from there," he told me.

It took me all day from Gene Jack's to Majuba. Maddie and John weren't there so I camped at their haystack and fed their cows, and then the next morning I took off for Salmon River. By that time the snow was deep and crusted because it had rained and snowed and then rained some more and froze. The crust of snow would support me most of the time but not the horses. I set out on foot leading one horse, which had to break trail by jumping on the snow and breaking through. When it got tired out, I used a different horse to break trail and put the lead horse at the back.

It was really slow going but the worst thing when leaving Maddie's was ice under the snow where the trail crossed the meadow. The horses refused to break through that ice. I didn't have gumboots and didn't know how deep the water was, so I took the horses clear around the whole meadow. The meadow was no more than one hundred yards across but it took me all day to get there. By evening I could see where I had camped the night before. I tied the horses up, fed them a bit of grain and camped right there.

I got up the next morning and went eight miles, as far as Ulkatcho Village, and spent the night there in one of those fallen-down cabins. When I took off from Ulkatcho in the morning it was snowing, and then it turned to rain. It snowed some more, then cleared up, and the temperature dropped like crazy. As we headed up the Salmon River Hill, the alders were crisscrossed over the trail, and in the deep snow the horses couldn't go under them. I had an axe but it was so dull it might as well have been a stone axe. When it turned cold about 2:30 p.m. I got a bit scared because the horses were plumb played out and wouldn't walk any farther. I figured it was either me or the horses, so I dropped the halter rope and left them standing there. It was only about four miles to Salmon River and my parka was soaking wet. I had only taken enough grub for three days because I thought I'd be out in no time, and I was famished.

I got to Pat Sill's place at Salmon River about eight or nine o'clock that night. When I knocked on the door they were startled. I said I was in a bit of trouble and told them about the horses I'd left behind me on the other side of the hill. Martin Toney was there and he said, "You better come in, there's nothing we can do tonight." I guess I was so full of relief and exhaustion that when I stepped into the warm cabin, I passed out.

That night the temperature dropped to thirty-five below. If I had stayed with the horses, I doubt if I would have survived. I'd still be out there somewhere.

The following morning they loaned me a horse and a couple guys rode back with me to get my horses. The horses were standing on the trail in the same place I had left them. They hadn't tried to turn around or anything: they just stayed right there. It was around noon when we got there. When they heard the other horses they nickered. It made a difference for my horses to follow a broken trail to Salmon River. We were all back at Pat Sill's place through the deep snow within an hour and a half. I spent one more night there before heading up the road to Corky's place, where I wintered the horses.

Later on I got a good compliment from Lester Dorsey. He had been hunting out at Tanya Lake when the Big Snow came, and his hunters had to be flown out by helicopter. With great difficulty he managed to get his horses home through the deep snow. One day I was helping Lester at Beaver Creek and when he heard I'd come out with eleven horses he said, "This was the winter that separated the men from the boys."

The following spring there were horse bones all around the country. When a horse is pushing snow with his chest it doesn't take him long to peter out. For many years after the Big Snow, every time it started to snow I got claustrophobic.

GETTING OL' YELLER UNSTUCK

CORKY: You couldn't see Ol' Yeller's muffler sticking out of the snow until the following spring. When we finally spotted it and located Big Fred's favourite tractor, we thought we'd hit pay dirt. Finding it was one thing but getting it out was another story. Ol' Yeller was in the middle of newly rotovated ground, and by the time all the snow had melted the field had turned into a bog hole. It was impossible to reach the tractor to pull it out. Big Fred and I decided the only chance we had would be to wait until the ground was frozen the following winter, and to call on the expertise of Lester Dorsey.

By March, Anahim Lake country is starting to awaken from the frozen deep. That's when temperatures warm up above freezing during the day, and still dip pretty solidly below zero at night. Ol' Yeller had been sitting in Bill Lehman Meadow for a year and a half, and Big Fred and I knew this was the time to rescue it if we were going to do something about it. Somehow we had to lift it up, get some logs under it and pull it out. I talked this over with Lester Dorsey and he said he would come look the situation over. I figured if anybody could do it, it would be Lester.

We took our snow machines down to see what tools and equipment we would need. Our first chore was to chop all the ice that was surrounding the tractor and to dig out the rotovator that was still attached

THE FAITHFUL TRACTOR, STUCK IN THE MUD.

to it. The wheels were frozen in the ground and we would have to chop them out with heavy axes, sledgehammers and wedges.

We had our chainsaws with us so we started cutting the ice into blocks to the waterline and chipped them out to create a three-foot-wide perimeter around Ol' Yeller and the rotovator. This was heavy, hard work and took us all day. We were wet and cold as the sun was beginning to go down and we headed home to Muskeg.

We made a list of tools we'd need for the next day including axes, crowbars, digging bars, cables, big hammers, a small propane tank, a tiger torch, several tarps, logging chains, ropes, a barrel of clean diesel, jacks, stove pipe, a siphon hose to drain off the old diesel from the tractor and some buckets. We also brought along a dozen two-by-twelve-inch cedar planks in case we needed to make a bridge, and loaded everything onto the wagon. Big Fred towed the wagon behind our four-wheel-drive tractor, which we planned to use to pull Ol' Yeller out of the swamp, and Lester and I took our snow machines.

We couldn't get the tractor and wagon any closer than a hundred feet of Ol' Yeller, so we parked them on high ground and carried everything over from there. The first thing we did was finish chopping the

THAWING OUT OL' YELLER.

ice around the tractor, which gave us some working room to get underneath the axles and the oil pan, still frozen in the mud.

We took some canvas tarps and built a tent over Ol' Yeller and the rotovator, shoved a piece of stovepipe under the canvas, placed the tiger torch at the end of the stovepipe and turned it on to a very low heat. As the snow and ice melted I had to dip out the water.

Meanwhile, Lester and Big Fred cut some twenty-five-foot-long Jack pine poles and skidded them out on the ice with the snow machines. As the sun went down we headed back to Muskeg on the snow machines for the night, with high hopes that we would free Ol' Yeller from the bog hole the next day.

The next morning Lester explained over breakfast his plan to free Ol' Yeller. When we got to the work site I relit the tiger torch. While we waited for it to warm up inside the tent we had made for the tractor, we cut holes in the ice for the legs of the A-frame built with the Jack pine poles and ran a cable out to the four-wheel-drive tractor. Then we ran a chain from the A-frame to the axles of Ol' Yeller front and back.

By now the ice and snow had melted sufficiently underneath the tractor and from around the wheels, and we cut off the tiger torch and

LESTER DORSEY GRINS IN THE FACE OF THE IMPOSSIBLE.

removed the tent. Big Fred got on the four-wheel-drive tractor, put it in low gear and crept slowly forward. As the cable and chains began to tighten, the A-frame began to lift up, and Ol' Yeller followed. Once Ol' Yeller and the rototiller cleared the ice, Big Fred went forward with the tractor about twenty feet. Then we put the two-by-twelve-inch cedar planks down like a platform and Fred backed up to loosen the cable and chains and set Ol' Yeller on the wooden planks.

We did this three more times before we got the tractor and rotovator to high ground. Lester had a big grin on his face. He seemed to get a kick out of doing something that seemed impossible. Big Fred was really happy too. He had his favourite tractor back. Now we had to find out if Ol' Yeller was damaged or would ever run again. That would have to wait until the next day. The sun was setting, but Ol' Yeller was sitting high and dry. It was pitch black by the time we got back to Muskeg, but Jeanine had cooked a good supper for us, which we ate before we fell into bed.

The following morning we made a tent around Ol' Yeller with heavy tarps and warmed it up with the tiger torch and stovepipe set-up for a couple of hours before trying to get her started. Meanwhile we drained off all the old oil into a bucket, and replaced it with new, lighter oil. Then we siphoned all the old diesel fuel out of the tank and replaced it with fresh diesel. We also changed the oil filter, put in new antifreeze and a new fuel filter and primed the pump. We had done everything we could think of as Big Fred climbed behind the steering wheel of Ol' Yeller. The moment of truth had arrived. I climbed onto the four-wheel-drive tractor and went forward to tighten the chain. Then I changed into a higher gear and got some speed up.

Once we got going pretty good, Fred let out the clutch and she started belching black smoke. This was a good sign because it meant the motor was turning over and wasn't frozen down. The smoke got thicker and blacker and chunks of black rust came out of the smoke stack as I gave my tractor a little more gas. Fred let the clutch out a bit more and lo and behold, Ol' Yeller started.

I really didn't expect this and neither did Fred. He had a big grin on his face and shook his head in disbelief. Lester said, "Well now, boys, I could sure use a stiff drink right about now."

Fred nodded, "Me too."

I told them both I had something that would make them smile. I walked over to my snow machine and fetched a brand new bottle of Wild Turkey from out of the console. We toasted Ol' Yeller. I told them I had a couple more Turkeys back home at Muskeg and we would do some more toasting that very night. It was a happy, happy day.

After we loaded the snow machines and all of the equipment and tools into the wagon, we kissed that bog hole goodbye. Lester took a shortcut to Muskeg on his snow machine while Big Fred and I went the long way around by the main road. I drove the four-wheel-drive tractor and Big Fred took Ol' Yeller. We got home just after dark and Lester was there. That night we killed a couple of Turkeys and had a celebration. It was something that I would remember for the rest of my life.

HAULING OL' YELLER TO MUSKEG

SAGE: There are a few stories of Ol' Yeller. Mike McDonough remembers giving the old tractor a unique ride from Anahim Lake to Muskeg behind his pickup truck one winter. Mike drove up from Kleena Kleene with his son, Jamie, and friend Pete Bookmyer in his old orange 1959 GMC pickup truck towing a car trailer for the tractor. Of course Corky warned them not to shut the tractor's motor off

CORKY AND MIKE MCDONOUGH.

because the starter was broken and they'd never be able to get it going again to back it off the trailer.

They got it loaded, chained it up so it couldn't go anywhere and left it running on the trailer like Corky told them. When they stopped to buy fuel in Anahim Lake before heading downriver to Muskeg, somebody warned them that Poison Lake Hill was really icy and that they ought to stop and chain up before starting up the long incline.

Mike says he thought, "Ah, I've got lots of weight on my back axle and I won't have any problem because I'd taken that old 1959 GMC places you should never go."

As they approached the hill, Mike poured on the gas and picked up speed. "We almost got to the top before I spun out on the ice," he says. "Then the truck and trailer, with the old yellow tractor still running, started sliding backwards down that hill."

They were faced with a dilemma. On one side the rock face went straight up, and on the other side the cliff dropped off straight down 150 feet or more. Mike says as they started sliding backwards, Pete told him, "I'm getting out right here," and he took Jamie with him.

"We only slid back about six feet," Mike says. "Sliding backwards on the ice—you put on your Indian seatbelt when that happens. That's where you grab the seat with your ass."

Fortunately the truck and trailer carrying Ol' Yeller were spared. Breathing many sighs of relief, they got the chains on and made it the rest of the way up that hill and all the way down to Muskeg without any further incident.

BOYS WILL BE BOYS

JEANINE: The guys often got into trouble because they wouldn't believe they couldn't do something. Usually, if somebody else could do it, they'd say, "I can do it! I'm fine, I know what I'm doing." And then the next thing you know there's little bits and pieces scattered all to hell. All the guys I knew up there were that way.

CORKY: Ol' Yeller caught on fire eventually and burned all the wiring off. We never did put a starter on it, but you pull it ten feet, and it would go every time. When Ol' Yeller caught on fire we got it out. We weren't so lucky when one of the four-wheel-drive Universal tractors caught on fire. It burned right up. Our son John drove Ol' Yeller a million miles. He was driving tractor when he was nine years old. He was so small, when he was driving Ol' Yeller he couldn't sit in the seat and reach the clutch with his foot, so he had to stand up. He was a remarkable kid. When he was young, you couldn't get shoes on him. He didn't like them. He would do a little dance at the top of the hill when somebody was coming, then he'd take off and run to the house.

He moved faster than anything I've ever seen. When we would come home from haying or something, John would get out of the truck and open the gate then run down the little path and race us to the house. Jeanine remembers him running down that hill full tilt, his bare feet and legs going like pistons.

REPLACING COWS AFTER THE BIG SNOW

CORKY: That year after the Big Snow, the Department of Indian Affairs hired George Reed to go out and buy some cattle to replace all the cows that had died. A lot of the cows were stuck way off in the toolies when the snow came and never made it home. Not many of the cows that were stranded made it through the winter. There was a big die-off. So George bought some nice Black Angus cows. Of course, once he unloaded them they didn't have a clue what to do. They didn't know the country.

Peter Alexis got a few, Chantyman got some, and the Jacks got some, but they weren't cow men. They just turned the sons of bitches out and if they lived, they lived, and if they didn't, well, that was just the way it was.

Anyway, the summer after the Big Snow, a cow came to my front gate. It was one of these polled Black Angus cows and I got to looking at it. What the hell was this cow doing over here? When I walked up, the cow didn't even acknowledge me. The first thing I did was look over her face, and then I took my hat off and waved it in front of it. She didn't flinch; she was as blind as a bat. She must have got hung up along the river where the bugs are by the millions. I let her in the gate and she was just skin and bone and in a lot of pain. The mosquitoes and blackflies had eaten all the meat off around her eyes. Her eyebrows were gone and she didn't have an ounce of fat on her. It was terrible. I couldn't do anything with her, she was too far gone. She lay down and died within four days.

BUILDING THE HOUSE AT MUSKEG

JEANINE: We started building our new house at Muskeg the summer after the Big Snow. Corky had always planned to build a house, and he wanted it on the high spot across the creek from the old cabin. I think he started talking to Mac Squinas even before we ever moved down there. We started getting the timber in as soon as the snow was off the ground in the spring of 1975. It took two years. We had to bring the logs in, peel them and dry them. Then, in the summer of 1976, Mac came and set up camp with all his family and some other folks, and they started building.

CORKY: There were a lot of people involved, but Mac was the main contractor. I paid him and he paid the others working with him. Eddie Sill was there with his family; there was Georgie and Bella Leon, and Dogan and Liza Leon. Liza babysat several babies at the same time. She had them in willow baskets hanging from the cabin ridge poles.

JEANINE: It was so nice to finally have a big new house to move into. By then I was really glad to move since the two kids were getting

BABIES IN BASKETS AT THE CONSTRUCTION SITE.

bigger. In the old cabin we didn't have a place for people to sleep when they dropped by. Everybody was always sleeping on the floor. It was really nice to have more space.

It was a beautiful two-storey house built with logs. There was a main floor and a sleeping loft where Dana and John slept. We had a real staircase, and Fred made a beautiful decoration underneath it. He took small poles that he split and used to cover the triangular space under the stairs. He coated them with Varathane so they stayed the most beautiful golden colour.

The house had a bathroom and running water—now *that* I could go for big time. Oh, water coming out of the tap! We went all out and dug a pit so we could have a septic system. We had a hot water heater but didn't run it very much because it took so much propane. I'd only turn it on once a week to do the washing and have real baths.

THE NEW HOUSE AT MUSKEG.

We had one bedroom on the bottom floor and the sleeping loft was fairly large. It was split between John's part and Dana's part, and at the top of the stairs we had the water tank so we could have gravity feed. Of course, filling that tank in the wintertime was a job and a half. Whenever I hear the words "Briggs & Stratton" I feel my blood pressure begin to rise. Will it start? What is wrong with you? What do you want Mommy to do to you today?

We had a black plastic pipe running from the pump in the creek up to the second storey. It went through a hole in the wall and filled the tank. You had to take the pipe down and drain it, then get the motor and put it on the sled and bring it back to the house. It was quite the rigmarole. You had to cut a hole in the ice to get to the water. Trying to keep two kids clean in winter was a challenge.

The house was so nice. From our bedroom window the first thing you saw in the morning was Anahim Peak, so you could check the weather. And when you were sitting in the living room or kitchen, we had two big windows that looked out the other way at the Itchas and Ilgatchuz Mountains. I never got tired of looking at those mountains because they always looked different. A different shade, a different shadow, a different green; they were never the same.

We put so much of ourselves into that home. First, there was Mac's work. You just don't see that kind of craftsmanship and log work anymore. It disappeared with the old people. Then we had the cedar boards for flooring, which Maurice Tuck cut on his little sawmill in Bella Coola. He also did the cedar shakes for the roof. Charlie Cook did most of the plumbing, and Corky picked up some cabinets from somewhere. They were highly inefficient, but they were cabinets. We also put a wood cookstove in the middle of the kitchen.

CORKY: We hired Andy Cahoose to build Fred a pretty nice cabin on the other side of the creek. There was enough room in there for a big family.

Once we got Fred's house built we left the old historic land-mark cabin for the Native people who came up from Salmon River, because it was about a day's travel from where they lived to our place. I fenced off some pasture for their horses so they could keep them there overnight. Anybody coming out of the Blackwater stayed there too. People like Peter Alexis, Bernie Biscuit and Goat Herder and his outfit from Entiako.

I told Joe Cahoose if he ever needed to put his horses there overnight to go ahead and do it, and said there was a cabin there for them. Did they ever appreciate that! Old Joe shook his head, probably wondering what I wanted because I let them have the old cabin for nothing. Then he got to know me and he'd talk to me and say, "Maybe horse outfit come tonight. Maybe next week."

"Yeah, go ahead," I'd say.

"Oh, thank you, thank you," he'd say. Then he'd go get a nice chunk of deer meat or moose meat and give it to me.

JEANINE: They appreciated it, I think, because a lot of the people back then were not so accepting of Native people. Once they realized that Corky saw them as people like everyone else, it was very meaningful to some of the old people that he would accept them as neighbours.

CASH MONEY ON THE TRAIL

JEANINE: Our nickname for one guy living way off in the bush was Goat Herder. He lived down by John and Maddie Jack in the Entiako

River country, fifty or sixty miles past our place, with his wife and five or six children who were all born in the bush. Despite having all those children, his wife was still the prettiest woman imaginable. But Goat Herder could have done a lot more. It seemed she worked really hard, and he didn't.

CORKY: At some point out there living in the wilderness, Goat Herder and his crew got religion. One day he got down and prayed, and what came out of the prayer was that he was supposed to leave some money for the next guy who came along. He put the money in a paper bag and put it on the trail. Eventually Bernie Wiersbitzky came along and found the money lying there in a bag, right in the middle of the trail. You couldn't miss it, he said.

Up the road somewhere he ran into Goat Herder and told him he found this money in the middle of the road. Goat Herder changed his mind and wanted the money back. But Bernie said, "I found it and I'm keeping it. You asked the Lord to send it down, and he sent it down and gave it to me. By God, I found it and it's mine."

PROPER HOSPITALITY

CORKY: I can't remember how many kids Goat Herder had—five, maybe six. I hated to see him coming because he'd stay a few days and his outfit would eat every bit of grub we had. Then he'd leave after the food was consumed and head on into town.

SAGE: Corky and Jeanine's daughter, Dana, remembers Goat Herder stopping by their place in Muskeg a couple of times a year.

"One time I got into bad trouble from my mom," she recalls. "Mom never got mad at us kids, so this was unusual. I looked out the window and could see Goat Herder and his whole pack of kids coming. I always freaked out when I'd see them because they'd eat everything but the wallpaper. They'd eat you out of house and home. Of course, my mother would open the door to anybody. They showed up quite unexpectedly that day and Mom had been making fried chicken for us. Those boys ate everything, and I only got one small piece of chicken. I threw a fit about it in front of everybody.

"Mom dragged me off into the bedroom and she spanked me and said, 'I don't know when those boys would have eaten last. You just shush. You had plenty on your plate. I don't ever want you to talk that way again.' Mom understood what being poor meant because she had come from extreme poverty herself. There was never a person she turned away. She didn't make a big deal of it."

HERONS AND MARMOTS AND GEESE, OH MY!

THE STORY OF THE BAD-LUCK BLUE HERON

JEANINE: The wildlife, animals and birds were a big part of our lives at Muskeg. Especially the birds. There were so many wonderful birds down there. Sandhill cranes would come through, and their sound would reverberate across the meadows and swamps.

One time we had a beautiful blue heron. He was so gorgeous. He fished there in the little pond where the creek slowed up right by the house. He picked fish all day until he got full. He didn't nest there. His mate must have been somewhere else.

But then Andy Cahoose almost shot it because of a superstition that the blue heron was a bird of death. He said, "If it flies over, it's a bad omen."

Corky and Fred were over working on Fred's cabin and I was up at our house. When they came home for lunch, Fred told me that Andy wanted to shoot the blue heron. I told Fred I didn't want that bird shot. I usually try to be sensitive to Native legends or superstitions. I never laughed at Fred because of his superstition of being scared of frogs. But I didn't want that bird shot. I was very clear about it. He was a special thing. I told Fred I would take all the "bad" that came from it, and Andy wasn't to shoot that blue heron.

CORKY: When Andy Cahoose was building the new house for Fred, he'd bring his whole family, and we'd go out into the forest and pick out the logs he wanted, then I'd haul them over with the tractor. They'd peel them and let them dry, and every few days he'd show up and catch up on the work. They lived close enough that they didn't camp at our

THE CREEK FLOWED THROUGH THE YARD AT MUSKEG. THE HERON POOL WAS BELOW THE FOOTBRIDGE ON THE LEFT.

place. Our house was nearly finished and we were living in it already when Andy started building Fred's cabin.

For whatever reason, the Native people didn't like the blue heron. Andy said the old people believed it was bad luck. They'd get a little edgy if it ever came around, especially if it went down the creek and started squawking. If it did that, they were sure somebody was going to die. That's what the old-timers believed, and Andy was one of them.

When we were back working on Fred's cabin, the blue heron flew up on the ridgepole of our house across the creek. We had the roof over the main part of the building but the roof over the porch wasn't up yet. So the bird landed on the ridgepole and stood there. Andy asked if he could shoot it, and I told him my little girl talked to that heron every day. It was her friend.

Andy persisted, so I told him, "You gotta do what you gotta do." I had a dilemma. He knew I didn't want him to shoot it, yet on the other hand I wasn't going to grab his gun out of his hand and start a big row. Anyway, he decided he was going to shoot the bird, so he got a bead on him and fired and missed him completely. The blue heron flew off unharmed. That was very unusual for Andy. Here you've got a guy who was born with a gun in his hand, who had hunted for years and lived off the land, hunted moose, deer, lynx and bear, and he misses a shot like that. It made me wonder about the spirit of that bird. Andy was a deadeye Dick with a gun.

ANDY CAHOOSE, FISHING WITH A GAFF.

WILLIE THE WEASEL

JEANINE: We had a weasel that came into the house to play. Weasels eat the mice and are actually nice little houseguests because they don't get into your food. They don't eat anything but live kill, and don't tear stuff up like squirrels do.

We called him Willie the Weasel and he was one of my favourite creatures. He used to play by going up the stairs, then leaping down to the couch and bouncing down to the floor. He would do that round and round. He loved it. It was his favourite thing to do. We all laughed and thought it was wonderful. Nobody bothered him, so he kept doing it. We got used to it.

At the end of the winter Willie left and we didn't see him again until the following winter. Then he showed up with another weasel that we dubbed Mrs. Willie the Weasel. We could have been totally wrong about this; it might have been his friend Charlie. The two of them stayed there for several months during the winter. They would frolic and play and run around.

They weren't living in the house. They would disappear for a day or two, then they'd pop back up. I think they were under the house somewhere, or maybe in the woodshed. One time we were putting up wood and stacking it in the shed and a little bitty weasel flushed out from under the woodpile. He was so funny. He was obviously scared to death because there were all these giants standing around him. Then he went, "Rrrrrrh!" and attacked my gumboot. It was the funniest thing. I slowly moved my foot so he could get around it and see his freedom. I thought, "By God, they're gutsy little critters."

PLAYING MANDOLIN FOR THE MARMOT

CORKY: One day a marmot came by when I was standing on the front porch playing my mandolin. I used to play my mandolin to tell the cows to come on up so I could build them a smudge. I did that every day in the summertime when the bugs were bad.

We didn't know what to think of the marmot or what was he doing there. We didn't even know what he was. He looked like a big ground-hog. He must have come across from the mountains on one side of the valley or the other, and he showed up at our door. Dana named him Columbus, because she said he must have been a great explorer to come so far. She was about three and a half or four years old then.

That marmot sat there on his butt and held his hands up when I played the mandolin. He liked that music. Somebody has a picture of me with my mandolin directing the cows and singing to them, and the marmot standing there. I think he moved under the house and lived there for a couple of years.

LONE GOOSE

JEANINE: In the spring we had a pond about one hundred yards from the house at Muskeg. It would pretty much dry up in the summer, but in the spring it was a good size. The geese would come in there and rest up for a couple of weeks on their flight north, then they'd move on to the bigger lakes once the ice melted off them. One day I was in the kitchen and I could see out the big window that two bald eagles were attacking a goose. They had it down and were killing it. This other goose kept trying to chase them off, but they were much bigger birds and way too much for him. He couldn't break into the attack. They killed and ate the goose they had down, then eventually flew away. A little bit later that other goose came back and he walked all around what was left of the body. Feathers and feet were about it. He sat down beside her remains and let out the most incredible sound. A terribly mournful lament. He did that for an hour, off and on. He'd walk around the spot; then he'd sit there and mourn and cry. He came back every day as long as the geese were there. Then when they started to pull out, he went with them. We didn't think much more of it except for the strange mourning lament.

The following spring we heard that same unmistakable sound. We looked out and there was the goose, right on the spot where his mate was killed. He walked around it and he sat down and did that same mournful honking sound. He came back three years in a row, lamenting his mate. He'd usually stand there for several hours, walking around and wailing. Then, after three years, we never saw him again. Maybe he died. I don't know what happened to him. We called him Lone Goose, and he was just one of those things that were a special part of our life.

CORKY: There were all the different kinds of ducks that landed there too. There was a black-headed tern that goes from the Arctic to Antarctica, halfway round the world and back. A wonderful group of birds moved through there.

I got a bird book and we counted something like eighty different species of birds that came through there. Anytime a new bird landed I got my bird book out to see if I could identify it. One of the most exciting sightings was the lesser bittern. The bitterns are patterned to live in the tall reeds and tall grass. They have striped markings that camouflage them so they're almost invisible. I don't think I would have seen him if I had not been walking right toward him until he felt he had to move. As long as he was standing still, his colouring blended perfectly with the tall wetland grass there.

One time I was walking over to a piece of land I wanted to look at to put in a ditch to drain the water, and I noticed all these crows coming in. This was the middle of the day, and they were flocking in from everywhere. So I got up in the bush to see what the heck it was. Two crows had landed out in the open on the meadow and this herd of crows gathered in the spruce trees around them. It seemed like the two birds were on trial. Had they done something wrong? I had no idea. It seemed like the other ones were mad at them, and they sat in the trees, squawking and raising hell. I had never seen anything like it. Hundreds and hundreds of crows. They stayed there about fifteen or twenty minutes. Then they all took a run at the two birds in the meadow. They never did hit them but they'd come and brush them a little bit with their wings as they came down. It seemed like they were punishing them for something, but I couldn't imagine what.

ON CREATIVE COMMUNICATION:
MUSKEG MOLLIE CALLING HANGING TREE

JEANINE: D'arcy Christensen was the flying fur buyer and he used to fly Sister Suzanne, the nursing sister, out to our place so she could give the kids their vaccinations and check up on them. One winter he flew our Christmas dinner in. He landed right in front of the house with a big turkey.

At that time we had a CB radio. Everybody down there had a CB, because it was the only way to communicate. You had to get on a particular channel and say things like, "Muskeg calling Hanging Tree… Muskeg calling Hanging Tree…" It was great. We couldn't reach Anahim Lake, so we relayed. Ken and Linda Karran were Hanging Tree. I was Muskeg Mollie.

I remember teaching Dana how to use it when she was just a little thing. I showed her how she could drag this box over and stand on it to reach the radio. I always used to leave it on the channel we used to talk back and forth, and I showed her how she could call on the phone for help if she was ever in the house by herself. She got so she could get up there and use the phone, but she never had to, thank God. A lot of times in the winter people would turn on the CB and listen to everybody. It was better than a soap opera.

We charged the CB battery off the tractor. We had a generator but we didn't use it very often because it was so hard to get fuel down there. We had a fridge and a hot water heater that ran off propane. We used the generator to start up tractors and stuff like that, and I'd run it once a week to do the laundry.

PINK JESUS

JEANINE: There was this guy, Steve Andruss, who bought the Bill Lehman Meadow from us. The Native people called him Squirrel Eater. He and his wife were hippies from Oregon who were into getting back to the land. One night Steve ended up at our house and he and Corky proceeded to get really drunk. Steve had a long dark beard and long dark hair, and looked exactly like those pictures of Jesus in the bible stories.

CORKY: We were moving stuff from Bill Lehman Meadow. We worked hard all day to get everything done and then headed down the road to Muskeg like a bunch of gypsies. Steve and I tied one on pretty good once we got back to our place, but it was too much for him. He got sick and he woke up puking. He had no clothes on so he grabbed a pink blanket that was in the house and ran out the door to barf. All he had on were gumboots and this raggedy pink blanket.

The cows got pretty upset with him running through the barnyard retching and making a hell of a racket. When one cow saw him, her knees turned to rubber, just like a man's do when he gets terrified, and she went down like she'd been shot. She tried to get up, but no way could she get back on her feet, because she had rubber legs. Then the whole outfit stampeded and knocked the fence down. The last we saw of them they were heading downriver to Peter Fuller's outfit at Ildash. Two of those cows shit for twenty-five feet when they saw that wild-looking son of a bitch coming out of the timber. They hadn't seen anything like that, or smelled anything like that before.

Somebody said he looked just like Jesus as he crossed the meadow with that pink blanket. After that he was known as Pink Jesus. Poor Steve. The Native people all gave the white people names. Mine was Nedo Yaz, which means small whiteman. I asked Fred why they called Steve Andruss Squirrel Eater. Fred said it was because that's what he lived on.

Steve and his wife didn't last but a year. They went through one winter. Leo and Vivian Hermsen eventually bought their place.

ALL IN A DAY'S WORK: THE ROTOVATING PROJECT

CORKY: I've got a picture of my son, John, and Eugene William standing in the door of Eugene's cabin in Nemiah Valley. Eugene was a great storyteller. Once you got to know him he'd turn loose and talk about some of that country. His dad, Old Sammy Bulyan, is the one who wore the sheet iron all around him. Somebody had shot at him, and this old boy was an old-time Native man. He wore an A-frame of heavy metal after this guy shot at him, and he walked with this thing everywhere he went, Eugene told me. It had some hand-holds where he could pick it up.

JOHN WITH EUGENE WILLIAM IN NEMIAH VALLEY.

John used to come with me and drive tractor. He was probably ten years old when we were in Nemiah. By the time he was in his early teens he could run any machine on the ranch. John helped me do everything. I told him to do something, he would do it.

We had to do our own mechanicking because there wasn't anybody for two hundred miles to come work on the outfit. Besides, we couldn't afford to hire someone to monkeywrench for us in the first place. Maybe we didn't even know what we were doing, but we had to fix it the best way we could. That was the only way.

I was doing some work for Voyne Purjue in Nemiah Valley, rotovating his fields with my tractor, when word got out to the Indian Department that I was breaking up land for the Native people. The Indians would hire me to do fifteen or twenty acres or something that they could afford, then we'd plant it for them to see what it would do. They all treated me good, too.

When the Indian Department guy came through, he asked if I'd be interested in ploughing up some Native land in other parts of the country. I told him for sure I would. I got to go from Nemiah Valley all the way to the Blackwater River doing work for the Indian Department. That's how I got acquainted with the Native people, and they saw I wasn't a shithead white guy.

CORKY TAKES HIS TWO UNIVERSAL
TRACTORS OVER THE DAVIDSON
BRIDGE INTO NEMIAH VALLEY.

I brought the first round baler and the first Brillion seeding machine into the country. The Brillion seeder is a special machine that packs the seed in the soil so it will germinate properly. The round baler completely changed the way we hayed because we could leave the bale in the field. With the square bales you're fighting them every day. The moisture comes up through the ground every night at Anahim Lake. Even when it's not raining you have to get those little square bales off the field or they'll get waterlogged and rot. The round baler helped you deal with this hay problem. All you had to do was harvest it. The Indian Department gave me a good-paying job where I got a paycheque. There wasn't much cash in the country in those days. Then I got my big Case tractor and twelve-foot rotovator that could do a lot of work in a small amount of time.

When we first got to Muskeg we got a little rotovator that would fit Ol' Yeller, our yellow tractor that came with the Holte oufit, and we ploughed everything up that was tillable. The wider you get with your rotovator, the bigger the tractor you've got to have to do the job. Ol' Yeller is what we started with. Then we bought two small Universal four-wheel-drive tractors made in the Romania. They cost about a third of what an American- or Canadian-made tractor would be. There was nothing fancy about them but they did the job. I put balloon tires on the front wheels. The wider the tire, the better it is in that soft ground. The wider, lighter, low-pressure tire keeps you from going down. On the back we put on caged wheels. You have to get a welder to put three little hooks on the rim of the back wheel to hold the cage. They use them in Texas out on the rice paddies. They are like a hamster wheel made out of pipe, and they give you a lot of buoyancy. With this thing you could get out in those big swamp meadows around Anahim Lake. I drove that thing where you couldn't take a horse.

Lester Dorsey about fell off his horse the first time he saw me out there. He came by to ride out to where I was working. He got out about ten feet and had to turn around and go back when his horse started bogging down in the soft mud. He said he never thought he'd see that.

Once we rotovated the Bill Lehman Meadow I sent a dirt sample off to Forestry and got it analyzed. It grew really good tame hay. The meadow dried off pretty quickly once the hay got growing, and of course we had the ditches where we could shut off and control the water so it would dry off. Those peat meadows are like a big sponge. If you dig too deep a ditch and take the water off too quickly, you'll ruin the piece of ground you're working on. All the water will run to the middle and the high ground will dry out and it won't be worth a shit.

It was a lot of trial and error trying to get the right machines and get the right wheels and tires. What we were looking for was buoyancy. When we got the Case tractor with the twelve-foot rotovator on it, one swath would take twelve feet. You could do more work in a couple of days with that tractor than you could with a regular tractor in two weeks.

THE GRAVE: EXHUMING THE BIG CASE TRACTOR

CORKY: I was starting to get lots of work with my big tractor. Down in the Klinaklini Valley, Pete Bookmyer got me to break up his place at Wheeler Bottom. Mike McDonough lived at Kleena Kleene and he learned to drive the tractor at Pete's place. That's where we got acquainted for the first time. I'd seen Mike on the road a few times before that, but in those days you didn't even know people at Nimpo Lake if you were from Anahim Lake. There were people living out there I'd never seen. They never come to town. Then I hired Mike to help me drive the tractor part-time.

Mike always claimed we took that son of a bitch places where no man should go with a machine. And he's right. I did all of Pete's meadow at Wheeler Bottom, then went up to the Blackwater where the Indian Department hired me to work up Peter Alexis's meadows. Then we came back and did Donn Irwin's lawn at his resort on Nimpo Lake, and Mike did a meadow for Bernie and Swede Gano along the McClinchy at Cariboo Flats before I took the tractor home to work up some of my own land at Muskeg.

MIKE MCDONOUGH DRINKS FROM A 40-POUNDER.

THE TRACTOR, MIRED IN DEEP.

SAGE: Several things conspired to create a perfect storm down at Muskeg Meadow. Dick Giles, a neighbour of Mike McDonough's at Kleena Kleene, owned a skidder that he had contracted with the Indian Department to do some land clearing work down at Salmon River. When a lightning strike touched off a forest fire near Lessard Lake, Forestry commandeered Dick's skidder to work on the fire. Meanwhile, Corky started working up a particularly swampy section of Muskeg Meadow that his daughter, Dana, dubbed Gloomy Point, because she figured it was the kind of place Eeyore would likely hang out. It was at Gloomy Point that Corky discovered a bottomless mire he never knew existed. He never guessed it possible that his big four-wheel-drive Case tractor would meet its match, but that's where it floundered. The more he tried to rock it out of the bog hole, the deeper it sank. When he finally gave up and went for help, the five-foot-diameter front wheels were buried below the surface of the meadow.

In Corky's case, going for help meant attracting more trouble. He learned that Dick Giles's skidder was just up the road on standby at the Lessard Lake forest fire and he managed to convince the operator

LIFTING THE TRACTOR TO GET IT ON THE LEVEL.

to bring the machine down to Muskeg to rescue his Case from the swamp. If anything could extricate his big tractor it would be the near-invincible skidder, built to traverse any terrain with its articulating chassis and powerful winch.

Wrong again. The skidder only made it partway across the Gloomy Point wetland before it, too, sank beyond moving in the swamp. Now Corky had two stranded machines to contend with, mired 150 feet apart in the bog hole.

About that time, Dick Giles stopped by Mike McDonough's place at Kleena Kleene to see if he wanted to accompany him up to Anahim Lake to visit Corky. "My skidder's up there," he told him. So both men piled into Dick's pickup and headed up the road. When they got to Anahim Lake, they stopped at Baxter's Café for pie and coffee before heading down to Muskeg, and that's when proprietor Don Baxter told them they didn't have far to go to find Corky, because he was in the motel next door.

"We banged on the door," Mike recalls, "and the first thing Corky told us was, 'You should see the grave!' He said his son-of-a-bitch big tractor was really stuck. Not only that, but so was Dick's skidder, mired in a bog hole on the Muskeg Meadow. That's when we started the saga."

DICK GILES AND MIKE MCDONOUGH. TRACTOR AND SKIDDER ARE FINALLY FREE AFTER A WEEK'S WORK.

CORKY: Gloomy Point is a particularly boggy part of Muskeg Meadow. There's a bay in the meadow where the mosquitoes come upon you unbelievable, and there's no bottom to the meadow there. It's got to be pretty bad ground if you can get a skidder stuck. The skidder is articulated, which means the whole machine is split into front and rear halves hinged together in the middle. All four wheels drive independently and that's how it turns. Usually a skidder can get out of almost any situation because they are so high off the ground and so powerful. The swamp must have been bottomless.

SAGE: Once the men got down to Muskeg, they assessed the situation. The skidder was buried right to the frame, so the first task was to free the skidder so it could be used to lift the tractor. Dick and Mike started cutting trees while Corky used his small four-wheel-drive tractors with the cage wheels to haul them over. "We cut the trees into six-foot chunks and pounded half a forest underneath the skidder before we were done," Mike recalls. He credits Dick with being a hell of a machine operator. "He'd get one corner of that skidder up and we'd just start stuffing trees under the wheel. Then he'd let it down and manoeuvre it around and get another wheel up."

Finally they got the skidder up on the level and started cutting corduroy across the meadow to the tractor. "We placed ten-foot lengths of logs side by side to give us a running surface on top of the meadow," Mike says. "Then we moved the corduroy forward as the skidder advanced until we finally had the skidder backed up to the tractor."

Even with the skidder's bull hook attached to the tractor, there was no way they could pull the tractor forward because it was sunk so deep in the mud.

"It had to be lifted," Mike explains. "We used Corky's little tractors with the cage wheels to help pull. That's how we exhumed the tractor."

Mike says when he got into the Case tractor to start it once everything was lined up to lift it out of the bog hole, the seat was at such an angle he couldn't sit down. "I stood on the clutch."

Mike praises the Case tractor as a workhorse. "Not only was it four-wheel-drive, but you could steer both the front and the back wheels. It didn't articulate like a skidder, but you could turn the front wheels one way and the hind wheels the other way, and turn that sucker on a dime. It was like a crab. You could crab it over and get into places no other machine could get into."

CORKY: It took a full week to get the tractor out. We had to put a sump pump in the hole to suck the water out so the fan wouldn't be whipping the water, and we had to keep the pump going constantly to get the water levelled off.

If conditions were right you could put a lot of acreage on with that bugger in a couple of days. God almighty, you could pull into a willow patch six or eight feet high and whack 'er down. Just chew it up and spit 'er out. Let it dry, then hit it again. It was looking pretty good down there by the time we pulled out of the country.

People are fooled by the first look of this wild-ass lovely country, but try making a living on the son of a bitch. You'll find out pretty quick you have to follow the good with the bad.

LATER YEARS

LEISURE TIME AND THE IMPORTANCE OF BOOKS

JEANINE: Almost everybody I knew out there in Anahim Lake read books. A lot of books were traded back and forth and one of the premier gifts you could bring someone was a book or two.

One summer Corky and I camped out with Bob Cohen and Francie Wilmeth at Knot Lakes for a couple of weeks. We flew out with Floyd Vaughan and one of Bob's friends, Brian Castner. We had a full camp set up with a little rubber boat. I went to get a book and looked up and realized we were all reading. I laughed. You've always got downtime in the summer and Bob and Francie always took books with them into the mountains.

The water in Knot Lakes was icy cold, but Dana would go into the water and play and didn't think anything of it. The only place around the lake that was a good campsite was also right in the middle of the bear trail. The tracks were deep from the grizzlies coming down that trail, because they always stepped in the same places. Their footprints had worn these deep holes down into the ground. Corky, Bob, Brian and John went off around the lake in the rubber boat and saw a big sow grizzly and her cub sitting on a point of land along the lake somewhere.

CORKY: We were in their territory. Come on in! We saw trees with big scratch marks high up there. That's how the big bears mark their territory. The higher up the tree they can reach with their claws, the more fear they can put into the smaller bears.

FRANCIE

JEANINE: I met Francie Wilmeth shortly after we bought Corkscrew Ranch from D'Arcy. She always looked a lot younger than she was, and

FRANCIE WITH HER YOUNG SON, PATRICK.

always wore her hair in pigtails. Her dad was the archaeologist Roscoe
Wilmeth, who did a few digs around Anahim and Abuntlet lakes and
along the Dean River, and Francie had come with him. We already
knew Bob Cohen quite well and I only met Francie a time or two
before she and Bob seriously got together. They came down to Lessard
Lake and spent a couple of days with us the winter we were there, but at
that time I wouldn't say I knew her well. Bob and Francie seemed like
the most mismatched pair of people I'd ever seen, but it worked out. It
worked out for a long time, then it didn't work out anymore.

I met Francie for real when Doug Archibald was having a branding party at the C2 Ranch up the Morrison Meadow Road and everybody was there. Eventually the branding party turned into a marathon
three-day drunk, and at some point I'd had enough. I was sober and left
the bunkhouse where everybody was partying and went over to Doug's
house. Francie was there, sitting in front of the fireplace reading a book.
That struck me as the best thing I could think of doing. So I looked
around and got a book and we sat in front of that fireplace and we read
and we talked. Then we'd read a while, then we'd talk. I think it was like

FRANCIE, PATRICK AND JOHN AT SEA.

the convergence of souls. It was really fun because Francie and I talked about things that most people in that country rarely talked about. Books, literature, a more literate world. It was something we both cared about.

After that it felt like she was a dear friend. She was my best friend during that time, and she is still a dear friend to this day.

Francie and Bob had a little cabin right in Anahim Lake where they stayed when they weren't on the trapline. It was down the hill from where Cam and Louise Moxon lived. Good God, what a party house. Everybody in the country has been there at one time or other. When we'd go to town (Anahim Lake), that's where we'd stay.

The Native people didn't read, but they played cards for hours and hours, and they gossiped and talked about other people. They played crib and just about any kind of card game. We introduced Fred and Daisy to canasta and they loved it. Fred and his son, John Lawrence, would come over to play.

We amused ourselves by creating our own entertainment. Reading books was really important to us. I would read whatever I could get my hands on, and I read to our kids from the time they were old enough to listen. We lived a little more in the now than what people seem to do today. We were more aware of the land and animals. It seems like people today live in a split world, only paying half attention.

DANA WILLIAMS: All my mother did was read to us. We didn't have TV, and only had radio occasionally. We didn't have electricity in the house, so reading was how we entertained ourselves.

Mother read very advanced books to us. That was our nightly activity. She read to us for an hour or two every evening, for our entire childhood. John and I were both fully comprehensive readers by the time we were four years old and could read books back to front. That comes from having no exposure to anything else. There was a lot of creativity and lots of games and imaginary stuff. Mom and Dad played with us a lot because you had to do something to pass the time.

NOT A LOT OF WOMEN OUT THERE

JEANINE: There weren't a lot of women out there. When we first moved to Muskeg, there was Sandy Zigler up at Rainbow Lake, and Linda Karran a couple miles away along the Dean River, and Francie. Louise Moxon was a good friend in Anahim Lake, and so was Goldy Reed. And there was Hazel Mars, the telephone operator. Those were the only women I was really friends with. Mickey Dorsey had already moved out and had pretty well left the outfit by that time. I knew the Native women but I wouldn't say we were close friends. Fred's wife, Daisy, was the exception. Fred's outfit was definitely like family.

Corky was always out zooming around doing things, and I was at home most of the time. He met twenty times more people than I did, and knew a lot of people that I never came in contact with. I knew of them, whereas he really knew them.

THE LEGENDARY LESTER DORSEY

CORKY: We spent lots of time with Lester Dorsey because he used the old Clark Place that was down the Dean River from Ken Karran and Andy Cahoose. He set up a hunting camp down in there because there

LESTER DORSEY, LEFT, WITH WOODY WOODWARD.

was a cabin and a creek and grass to graze his horses. It was an open spot where the bugs weren't too bad.

He was a tough man, let me tell you. We used to winter his team all the time. I told him, "Just bring your horses over there, and we'll feed them for you." He paid us for the hay. I didn't charge him much. He appreciated that because it meant he had a team down the river where he could use them.

JEANINE: Lester could really live in that country. He epitomized that country more than anybody else. He was one of those Americans who came up in the 1920s, several years before Pan Phillips and Rich Hobson moved in. He ranched in different places all over the country and was a great horseman. He raised beautiful horses. You could see a horse out there and you knew that was a Lester Dorsey horse. They had a fine trim look to them, just beautiful. Usually a light or a dark bay. Then he had his workhorses, too.

DANA POSES WITH LESTER.

Lester knew everybody and it was because of him that we learned about the country. We got along with him well. Not everybody did, but we did. Oh Lord, they had feuds out there that would have done the Hatfields and the McCoys proud. And most of them were over nothing.

MAN VS. BEAST: THE HIPPY AND THE HOG

CORKY: One year when we were living at Muskeg, some hippies moved into Lessard Lake. They had two hogs but the bear got one, and over the summer and fall the remaining hog put on some size and growed up on them. Soon this big brute of an animal became more than the hippies could handle. Mac Squinas told the hippies the best thing they could do would be to butcher that hog. "I told them it was going to die sometime anyway," he said. But the hippies didn't want to kill it because it had become their pet.

Everybody had CB radios in those days, and even Mac Squinas had one. Winter had come on, snow was on the ground and the lakes and creeks were frozen over when Mac Squinas called me up and said the hippies at Lessard Lake were pulling out of the country. They wanted to load their pig on a skimmer and haul it out of there behind their snow machine. He was wondering if Fred and I could come over the next day and give them a hand loading the pig on the skimmer—a freight toboggan shaped like half a banana pulled by a snow machine.

Bright and early the next morning, Fred and I got on our snow machines and headed over to Mac's place, then the three of us continued on the mile or so farther to Lessard Lake. We were wondering how the hell the hippies were planning to tie their two-hundred-pound hog onto the skimmer. A hog's body is shaped differently from most farm animals in that its head is smaller than its neck. It's hard to get a rope tied around its neck to purchase any hold on the animal. That's just the way he's built.

They had him in a little horse-breaking corral there. Anyway, we roped the pig and I handed the line to one of the hippies to see what he would do. The hog took off and the other hippy jumped in there to help him hold on, and the pig started dragging both of them around inside the little round horse corral. We were cracking up, laughing to beat hell.

Finally we all went in and grabbed the hog, hogtied him front and back and put him in the skimmer. That hog was just screaming bloody-ass murder. Then I asked the hippy how he planned to tie him in the skimmer. I warned him that this hog wasn't going to stand up with a smile on his face all the way to Anahim Lake. "You're going to have to tie this son of a bitch up, and tie him good."

So the hippy got in there and started tying ropes onto the pig, trying to lash him to the skimmer. He got it tied fairly tight, but he used the wrong kind of hitches and the hog wasn't secured very good. When they got out on the lake right in front of the cabin, they hit a rough patch of ice and the hog started to get loose.

One problem was that the track from the snow machine was kicking all kinds of snow into the hog's face, and this got him agitated. The hog started raising old Billy Hell and fell out of the skimmer. He skidded across the ice, but luckily he was still hogtied so he couldn't get any traction to get back on his feet. Finally we got another rope on that son of a bitch and loaded him back on the skimmer. This time we turned him around with his head facing backwards and hogtied his head securely to the crosspiece on the skimmer so at least the snow wasn't hitting him in the face, and we put a tarp over him.

The hippies took off again, and we followed behind in our snow machines, watching. The hog still wasn't liking this worth a shit, and was still struggling. They stopped again and I told the hippies they needed to tie the hog in tighter and sit down on him while he was being towed. Then I suggested maybe the best thing they could do for themselves and that hog was to shoot the son of a bitch. Butcher him right there. After all, it wouldn't be long until he weighed four hundred pounds. But the hippies were serious about keeping him for a pet. They even called him "The Pet."

They took off with the hog screaming bloody murder. You could hear his voice echoing off the mountains. It was the damnedest sound you ever heard. One hippy driving the snow machine and the other sitting on top of the pig facing backwards, with his hair flapping in the breeze as they headed up the trail toward Anahim Lake.

We followed them all the way across Lessard Lake, about two miles from where they set off, and down to Bill Lehman Meadow. It looked like they were getting along all right, and the trail they were taking was the only trail that everyone used, so if they needed more help we knew somebody would come along. Fred and I went back home.

Everybody in that country had a snow machine and we knew people came along that trail every day. We were wondering who might run into that hog and those hippies. Well, it turned out to be Lester.

I was out on the porch and could hear Lester coming on his snow machine. It had a distinct sound to it, and you knew it was him when you heard that sound. He had a different kind of muffler that went pop, pop, pop. Fred heard it too and said, "Lester's coming." He said Lester was bound to have run into them guys.

In wintertime Lester wore women's pantyhose on his head because he said it kept him warm. The two legs would hang down on either side of his head and he had knots tied in each one to shorten them a bit. When he rode his snow machine the wind would hit them and they would flop around. Holy jumpin' Jesus H. Christ, he looked like a crazy man.

He pulled up in front of the house and I said, "How you doing, Lester?" And he said, "Have you got any whisky?"

I said, "Yeah, we could probably do that."

Lester looked like some kind of demented rabbit with them long things hanging off of him. Then he said, "You know, you can see anything nowadays. Would you guys think I was out of my rabbit-ass mind if I told you I just saw a hippy ridin' a hog backwards down the trail?"

His eyes were big, and you knew you were talking to a man who had seen it all, but who had just seen something he had never seen before. I said, "Come on in, Lester, you better have something to eat."

We laughed about how those hippies had come to live off the fat of the land of Anahim Lake and were moving back to warmer country and taking their hog with them.

Years later I told this story at a cowboy poetry festival in Kamloops. This guy came up to me and he couldn't quit laughing. He said that was the best story to ever come out of Anahim Lake.

BAD BUGS

CORKY: We had horses in the beginning, but there were so many flies and bugs in that country, it was hard on them. Maddie Jack had come over soon after we bought the Corkscrew Place. She and her husband, John, lived downriver at a place they called Majuba, on the other side of Ulkatcho Village. They cut hay there and raised a few cows. We inherited an old team horse from D'Arcy when we bought the ranch and Maddie needed a horse to team up with another horse to pull her wagon. She came over and bought this big white horse, because she knew the animal. She said, "I'll pay you in the fall." So I let her take it, and she did pay me in the fall. I didn't have any particular use for one team horse.

Once we moved to Muskeg we were more into the farming, but we usually had horses around because people were always leaving them there. It was a great dropping-off spot. If somebody didn't have hay they'd leave their stock for me to feed, and I'd charge them by the month.

We had two or three horses all the time, just something we could get on and go. But there was really no place to ride them. The goddamn bugs would eat them up alive. I felt sorry for the poor sons of bitches.

JEANINE: We used to make smudges in the evening for both the cows and the horses: they could stand in the smoke and it would keep the bugs down. Then we got into spraying bug repellent. Those bugs would run the fat off them.

CORKY: In the heat of the day when the flies started coming out, I'd build the smudge by my house—Bob Cohen showed me how to make a green willow smudge with dry cow shit or horse shit. We did two or three smudges and the cows would come up and all percolate through there.

The worst bugs I ever saw were over at Ken Karran's. Betty Creek runs right through his place and those mean, biting sons of bitches just lay in there. With my round baler I started to get all my hay up a lot sooner so then I'd often come help Ken bale his up too. One day I heard this scream. I'd been hearing it on and off and I stopped my machine, turfed the bale out and started listening. The tractor was idling, and I heard it again, a horrifying scream. It was the horses being eaten up by the blackflies, mosquitoes, deer flies and bulldogs. Every biting son of a bitch known to man was down there. The bugs didn't bother me on the tractor because there was dust coming off of that round baler, but if you got down where these horses were, you could rub them under the belly and your hand would come out bloody red.

These horses would work their way through these little short willows at full speed and the whack from those willows would knock the bugs off them. But the bugs would get so bad the horses would scream. I couldn't believe it—it sounded like a death scream from hell, the wildest scream I ever heard. The horses were screaming on the dead run, and after they got through screaming, the whole outfit went out into the Dean River. The whole bloody outfit of eight or nine horses, right out to the deepest part of the river, with just their noses sticking out like hippos.

That's why the Dean River is the best fishing outfit in the world: nobody goes down there. They can't stand the bugs. Maurice Tuck said the bulldog flies beat him off his boat. At certain times of year the Dean River is so full of bugs, it's a blood bath.

JEANINE: Back at the ranch, we strung a cable between posts and wrapped it in rags soaked in bug dope. The cows would come and rub their backs along it to get that stuff on them. Oh man, we tried everything. The wild horses on the range got away from the bugs by going up to higher, drier land where there might be a breeze. But these poor buggers that were stuck in the swamps along the river really suffered.

CORKY: We humans suffered too. We had to sleep under mosquito nets to get away from the little bloodsuckers. One time Fred and I were working outside at Gloomy Point on Muskeg Meadow and we happened to look up and see this cloud hovering in the trees. It didn't look right, and it was moving pretty fast. Fred said, "That's goddamn mosquitoes." Then they hit us. The year after the Big Snow the mosquitoes were the worst because there was so much water in the bush for them to breed in. Every pothole was full of water and mosquito larvae. Dogan Leon said, "Cogie, you go down Beel Lehman Meadow, the mosquitoes they come off just like smoke."

OLLIE MOODY.

HOG ATTACK WHILE ROTOVATING
IN THE BLACKWATER

CORKY: I was rotovating land in the Blackwater for Pan Phillips and Peter Alexis, and was staying in a little cabin next door to George Chantyman at Peter Alexis's place. Peter had these two big hogs, and he just loved those pigs. He had them running loose. Then one day they broke into my cabin with me in it. They knocked the door off and came right in and tipped over the bed I was lying in.

OLLIE MOODY: I was living on Wes Carter's ranch over by Sleepy Hollow, close to Pan Phillips's place, and I would visit Peter fairly often. These pigs were three or four hundred pounds. Big white buggers. Corky was back there rotovating and the little cabin he was staying in was right across from George Chantyman and Cellia Alexis's place. Corky was in his cabin snoozing away when these pigs decided they were hungry and were going to help themselves to the grits in Corky's outfit. They busted the door down and came right in. Corky figured a bear had him right by the ass. He was on his bunk firing everything at them he could find. He couldn't find his six-shooter or he'd have shot them dead. George said, "Jeez, sure lots of noise go on over there. That little Texan sure jump around." I sure laughed at George.

CORKY: Those sons-of-bitchin' pigs knocked the stovepipe down and ran over it, and it broke into a thousand pieces. Then one of those hogs ran his head inside a big sack I had used for my groceries and got his head stuck in there. Then that son of a bitch did go nuts. He backed up, ran forward, hit the wall and kept going. The whole house shook. I was grabbing for my gun and Chantyman heard the excitement and came over. He grabbed a double battery cable off the wall and started beating the hell out of them. He was yelling, "I hate the pik! I hate the goddamn pik! Kill him! Kill him!"

We finally got those hogs outside and assessed the damage. It looked like somebody'd thrown a bomb in there. Stovepipe was all over the floor, trampled flat. The stove was knocked over. Lucky it wasn't going. They broke all my eggs and my food was scattered all over the floor. The whole thing happened in thirty seconds but seemed like an eternity. It took me a whole day to clean it up. Chantyman told me, "Peter won't let me kill that pik."

Chantyman said the pigs would climb up and knock down the harnesses and saddles they hung up under the eaves of the house. "The pigs would get in there, stand up and pull the saddles down, and start eating on them. They just ruined it," he told me. He hated those hogs because he had to repair all the damage they created. When they tore the fences down, he had to put them back up.

OLLIE: Later in the winter I had to go up to Irene Lake to pick up a horse, so I rode through Peter Alexis's place on a saddle horse. No one was home. Peter and Minnie had taken off and left the pigs behind. They didn't feed them; they let them fend for themselves. The pigs had rotovated that frozen ground right up. Corky couldn't have done a better job with his tractor. They rolled the big boulders and cleaned the willow brush right out around Peter's house.

CORKY: When the water went down in the Blackwater River, those pigs started digging holes in the riverbank and making tunnels underground. That got them down where the white root-tips of the plants start to show and they'd eat them right up. You wouldn't believe it unless you saw it. They slept in those holes in the wintertime like bears. Chantyman took me down there and showed me the holes. He said that's where those sons of bitches live. They've got a happy, happy home in there with all those roots hanging down.

OLLIE: Peter got those pigs as wieners in Vanderhoof and built a pen for them at first. This worked fine until they were six months old. Then they wanted to get to the other side where all the grass was, and they tore the fence down. Peter never did try and put them back in the pen after that.

When I came through with my saddle horse that winter, the pigs had tunnelled into the haystack and had a little place where they slept. I walked by there in the morning to go catch my horse, and there had been a fresh drop of snow in the night and the snow had slid down the haystack and covered the hole. As I walked by the stack, the pigs heard me and came boring out of that snow and scared the living shit out of me. The first thing I thought was bear. Then I turned around and saw the pigs right there.

Anytime Peter went to Anahim Lake with the horses and wagon after that, the pigs followed right along. No way were they going to be left behind after that one winter. Peter had those pigs a year and a half. When he got rid of them they were a good four hundred pounds. I was with him when he came to town with the pigs that last time. We were travelling between Beaver Creek and Ken Karran's place, and Ken came out of his driveway and we stopped to talk to him. The pigs were travelling kind of slow behind the wagon, but they eventually caught up with us. Ken says, "Damn, Peter, you got pigs."

He says, "Yeah, I got pigs."

"Well, I've been looking all over to buy a pig. I want to butcher a pig, I'm getting tired of moose meat and beef."

Peter says, "I'll sell you one."

Ken asked him, "How much?"

Peter says, "I'll sell you one, but I don't know what I'm going to do with the other one."

Ken says, "I'll buy both of them. The kids will want a pig too." So he bought them both. You know what he paid for those two pigs? Three hundred bucks. One hundred and fifty dollars apiece. And it was good meat too. Grass-fed and they weren't overly fat. They were just big.

THE COW BRUTE FROM HELL

CORKY: One morning Big Fred came over from his cabin to have his morning coffee. While we were talking over what projects we needed to do that day, the dogs started barking up a storm at the front gate. We thought someone was coming, so we walked across the bridge over the creek to see what was happening. The dogs were really haired up and I got to thinking it might be a bear. I went back to the house and got my old 30-06 just in case. As we got closer to the gate we saw that it was a cow standing there. I called the dogs off and went to see if she was branded. She was, but it wasn't any brand I knew from our part of the country. I asked Fred if he knew the brand, but he didn't. What stood out about that cow was that she had a really big head for her body and she also had great big pointed horns that curved out in front of her face. She had a mean eye and a nasty attitude. She was the kind of cow you don't want in your outfit.

She began to paw the dirt and blow snot and the hair stood up on her back like a mad dog. This cow was bad news, so we decided to sic the dogs on her and run her off and hope she'd go back home to where she had come from. The dogs put the run on her and sent her down the road, but not before she tried to hook them with her killer horns. We were amazed by how fast she was. She could turn on a dime. As the dogs kept hazing her down the road, she threw her tail over her back and let go a blue streak of shit on the run. I never knew a cow could hold that much manure. We had to take the dogs down to the creek and wash the cow dung off them. The outlaw cow was gone but I had a bad feeling we hadn't seen the last of her, and I was right. Fred and I both noticed that her right front toe was longer than her left, and she left a distinct track you could easily recognize.

We decided to check the fence line that day to see if any panels were down and needed repair. We gathered up the wire and pliers we would need for fence repair, and of course brought plenty of mosquito repellent to ward off the biting hordes. We hadn't gone far when we saw the unmistakable tracks of the rogue cow on the outside of the fence. She had tried to break through the fence in several places and had knocked the top log off. We went to work and repaired the fence. It took several days to make the rounds of the whole fence line and repair the damage she had done. We knew we had a chronic fence buster in our midst that would teach the other cows how to break fences—we couldn't have that. I told Lester Dorsey about it and he said, "Well now, I mean, a cow like that has to have a real bad experience. Like a case of lead poisoning."

At that moment we declared war on the Cow Brute from Hell.

She disappeared and we didn't see hide nor hair of her for several days. That all changed a few days later when Fred and I were working on a different part of the ranch and weren't at the house. The dogs started barking and Jeanine and John went out to see what the dogs were raising hell about. Lo and behold, there were about ten head of Angus cows that didn't belong to us. I had salt and mineral blocks out and Jeanine said they went right for them. It was right then that she saw the Cow Brute from Hell trying to break down the fence where it crossed the creek. It was a light fence, not of log construction, and she had managed to get her horns into it and was trying to break it down. Jeanine went running and started yelling and waving her arms. The other cows started moving off, but the Cow Brute was set on getting through that fence and paid no attention. Jeanine grabbed a stick, waded into the creek and started whacking that cow over the nose. It was an attitude adjustment, and she deserved it, too. She finally broke loose and headed back toward the outside of the fence where she'd gotten in. Jeanine herded the cows out and sent them up the trail toward Ildash.

When Fred and I got home that evening, Jeanine told us what happened and we went to look at the fence. She had ruined the gate so we had to fix that as well. We started to patrol the outside of the fence every few days looking for tracks, but we found none. We began to believe once again that she had pulled out of the country. She had vanished and we forgot about her. We set about getting ready for haying and getting our equipment into shape, and we needed some replacement parts for the machinery. Jeanine and the kids went to Williams Lake to get the parts and supplies, and Fred caught a ride with them to Anahim Lake because he wanted a few days off. He needed some time to blow off a little steam.

After a couple of weeks Fred was ready to come home, so he sent word by moccasin telegraph asking if we'd go to Anahim Lake to pick him up along with his family and their groceries and bring them back to Muskeg. I was at the ranch by myself working on the baling machine.

I finished what I was doing and decided to walk the fence line. I took my rifle just in case, called the dogs and started out.

The fence was in good shape and there were no tracks. Then about a hundred yards from the front gate the dogs got excited and I knew right then that the old hammer-headed outlaw cow was back. The gate had been hooked off its hinges and was torn up down by the creek. Then she had turned her wrath on the fence where it crossed the creek, which was the spot we had replaced before. She had charged the fence and got her head and horns through, but when she tried to back up, the top rail fell off and pinned her tight. That's how I found her. She couldn't move either way, backwards or forwards. This was the moment I was waiting for.

I got a rope around her horns and tied it to the fence to secure her front end. Then I roped her left rear leg and pulled it up tight so she couldn't kick me. I went to the shed and got my dehorning saw, and told that cow to brace herself for a life-changing experience. I returned and started sawing off her terrible hookers. She blew snot all over me and her eyes rolled back in her head. I sawed off each horn, leaving a couple of inches on each one so she wouldn't bleed to death, then I tied a bunch of tin cans to her tail with a piece of baler twine.

It was payback time for all the hours we had spent repairing fences while being chewed up by the mosquitoes. I was sending her a message the best way I could that she wasn't welcome back.

It was time to cut the ropes and let her go. To be on the safe side I jumped in my truck. Without her horns she backed up from the fence, but she was still on the fight. She was looking for someone to kill and that would have been me. Since I wasn't available, she took off down the road, the tin cans banging and clanging with every jump. I had taken Lester's advice and seen to it that she received an attitude adjustment she wouldn't forget. I never did see her after that.

FRED RUNS OVER DOGAN'S HEAD WITH THE TRACTOR

CORKY: We had just finished haying, after being blessed with a stretch of good weather, and had gotten our hay up in good shape, so I sent Big Fred and Dogan Leon to help Cless Pocket Ranch stook up their square bales.

DOGAN LEON, ONE TOUGH MAN.

It had rained for two days and then it cleared off. The bales were soggy and had to be stooked up or they would rot. Big Fred and Dogan took off early from Muskeg and got home late. I didn't see Big Fred until the following morning when he knocked on my door around six o'clock.

Lester Dorsey had spent the night with us and we were having breakfast. I told Fred to come on in and have something to eat, but he declined. He said he had some bad news, and was shaking his big head. This was a bad sign. I asked Fred what had happened and he said he had run over Dogan Leon the night before with the tractor. We sat in stunned silence. Lester said, "Well now, I mean, did he say he ran over Dogan with the tractor? That would kill most people. Is he dead?"

"No, he isn't dead," Fred replied. "I missed him with the big wheel. I ran over him with the little wheel. It mashed his head down into the mud. He was riding on the fender when I hit a bump, and it threw Dogan forward and under the front wheel. I had to back the tractor off his head."

I asked Fred if Dogan was okay, and he said he was, except he had a real bad headache and a tire track across his forehead. "He's out in his tent. He must be alive because he's snoring real loud."

We went out to the tent to see how Dogan was doing, and he was sound asleep and snoring without a care in the world. We could see a tire mark across his forehead. The soft mud probably saved his life.

Big Fred said, "Lucky we had all that rain."

Lester looked at the tire track across Dogan's head and said, "Well now, I mean, I have lived here a long time, but I have never seen anything like this." He broke out laughing and the rest of us did too. We couldn't help it. Dogan completely recovered. He was one tough man.

EDWARD AND SAMMY LEON'S BACKWOODS SURVIVAL

CORKY: One winter, Dogan and Liza Leon's two sons, Edward and Sammy, were up at Irene Lake doing some trapping. Irene Lake is about ten miles past Rainbow Lake on the road to the Blackwater, before you get to Eliguk Lake.

The brothers took turns running the trapline with their snow machines. One day on Sammy's turn to run the trapline, the weather turned really cold. He headed out on his snow machine and it got later and later. Finally Edward got concerned because Sammy hadn't shown up yet, and it was getting to be one of those fifty-below sons of bitches. So Edward figured he better go look for him. He got his snow machine going and followed Sammy's tracks in the snow, and he found him with his snow machine stuck in a creek. They both tried to get it out and ended up getting Edward's snow machine stuck too. They were both wet and it was close to fifty below and getting dark.

Fortunately there was a loose haystack in the meadow and Edward took his big long buck knife and cut a hole in the side of the stack. They dug it out and they both crawled in there for the night and survived. If they hadn't had a knife, no way could they have got through the outside of the stack, which was frozen hay at least a foot thick. The plug of ice has to come out of there. It's kind of like digging cement out.

OLLIE NUKALOW

SAGE: Ollie Nukalow was a legendary character in the West Chilcotin and Anahim Lake country. He was a half-brother to Big Fred's stepmother, Madeline Palmantier, who Baptiste Elkins married after Big Fred's mother died at Nazko. Ollie and Madeline's half-sister was Chiwid, or Lilly Skinner, the renowned Tsilhqot'in recluse who lived outside for fifty years, starting around the time Fred was born in 1933. Ollie, Madeline and Chiwid had another half-brother, Scotty Gregg, who lived at Kleena Kleene near Clearwater Lake. Ollie Nukalow had an unusual ability to endure solitude.

BIG FRED: In the fall time my dad took Ollie Nukalow to Knot Lake with a horse to go trapping and Ollie stayed there all winter. By July he never came out yet, and his sister (Madeline) worry about him. My dad worry too, I guess. Scotty Gregg was fixing his horseshoes, and he said he's going up to look for Ollie. I tell him, "I help you." We finish the horseshoes and we take off. We stay at Nimpo Lake one night. The next night we stay at Charlotte Lake with a Tsilhqot'in girl. We got up early. We got no meat. "A lot of deer track on the airport," the girl tell us. "Look on the sandbar by the lake." We went down there a little ways and shot one buck and we skin him out. That girl roast a hind leg for us, and we take it for lunch.

We took off pretty early. Get to Knot Lake about three o'clock. Scotty told me to make coffee. I just make fire when he come back. He said Ollie Nukalow still alive yet. "His track about a month ago," Scotty tell me. We had supper, then Scotty say, "I'm going to fire three shots." He fire three times. We listen. No answer.

We find out later that Ollie hear the shots. In his mind he figure it was a snowslide. Then he figured it can't be snowslide. It can't be two quick ones. After he went to bed he think about it. Then got up early and pack up. He got to the top of the hill and see our horse tracks and know somebody looking for him.

OLLIE NUKALOW, A LEGENDARY CHARACTER.

I don't know what he was doing. Sam Sulin used to trap down there too. He said Ollie sleep all the time. He said it snow two days ago and he said he see Ollie's snowshoes outside the little cabin. No tracks. No smoke. He figure he dead in there. He hollered. He sleep all the time. Just like a bear. He hunt all winter. Just enough to pay for his grub that he haul in. Sleep all the time.

CORKY: I heard Alfred Bryant talk about going down the Precipice in the cold weather. He said all of them got sleepy. When they got real cold they started to get sleepy. They go around and kind of shake one another, 'cause they ain't supposed to be going to sleep on the side of the hill like that in the wintertime. They get sleepy, that's what he said. He said you got to wake yourself up, or watch for it all the time. Some guys will take the cold better than others.

BIG FRED: Ollie Nukalow come out about five days later, and got back to Anahim Lake. At Knot Lake just sleep that's all. Just like a bear. He can't hunt, just sleep like a bear.

MOONING D'ARCY

CORKY: Haying at Muskeg using the round bale system was a lot quicker than putting up the square bales. One year the weather cooperated so we got our hay up early with everything in good shape. We always had some whisky stashed somewhere so I dug out a couple of jugs and me and Fred started to celebrate. We were sitting in the big field admiring our lush crop of green, round bales when we heard D'Arcy Christensen's plane approaching the ranch.

I'm not sure what possessed me, but I got Fred to help me push one of the round bales on its end so it would be a little higher than the others, and I jumped up on it and pulled my pants down and mooned D'Arcy as he went by. Anyway, D'Arcy went down and turned around and he came back and flew over us again. So I mooned him once more. This time you could see all the passengers cracking up. He did three or four loops.

Once in a while he'd call me on the CB saying he was bringing a group of sightseers down to have a look. He'd get his passengers all loaded in the plane and fly them down and I'd moon them, and they'd laugh.

Other pilots like Wayne Escott and Floyd Vaughan got in the act too. I'd hear them coming or they'd call me on the CB to see if I was home, and I'd give them the bare-ass salute. D'Arcy would fly over and do a little dip with his wings so I knew he had some people on board, and I'd go down and get on a bale and moon them. Often these passengers were from Vancouver or beyond, and they'd go back home and tell people they saw a naked man standing on a bale of hay. I'll bet you money D'Arcy was gambling a little bit on it.

ON THE DELICATE USE OF FIRE

CORKY: Fire was a method most of the ranchers around Anahim Lake used to get rid of old grass and improve the nutrition of their fields. Fred and I would regularly get up in the middle of the night and set the meadows on fire. We did it so often that apparently the Forest Service named a few fires after me. Cork One, Cork Two, all the way up to Cork Fifteen.

JEANINE: There was no other really good means of getting the old hay off those meadows in the springtime except by burning. Usually it went well, but every now and then the fire would get off the meadow and into the trees. That's when we really got in trouble.

We were all trained. Even the little ones were trained. When a Forest Service helicopter comes over, run and warn everybody. Then say you don't know a thing about it.

CORKY: Forestry wasn't too interested in letting us manage our land with fire. In fact, it became a big source of problems between me and the Forestry officials in Alexis Creek. They never could catch me, but they figured I was doing it. One of the Forestry guys singled me out and started giving me as much trouble as he possibly could, so I said, "I'll fix you sons of bitches. I'll get up in the middle of the night and set a fire, and by daylight I'll be in Kleena Kleene."

That was before they had the eye in the sky. A lot of people were doing the same thing, lighting a fire and then blaming somebody else for it. Somewhere on down the line everybody gets blamed for it, and nobody really knows who did it.

JEANINE: Burning meadows to get the old grass off is what they'd always done in that country. Back in those days there wasn't the amount of money or effort put into fighting fire like there is today. It was much more localized, like Forestry renting somebody's Cat. It was much later that they started putting the fire crews out. That was after people started making money harvesting the Jack pine.

WILLIAMS LAKE

JEANINE LEAVES THE CHILCOTIN AND MOVES TO WILLIAMS LAKE (1979–1986)

SAGE: The 1980s was a time of momentous change for Corky and Jeanine. They sold the Muskeg Ranch, the last remaining property from the original Holte estate, and Jeanine and the kids moved to Williams Lake. It wasn't a change that any of them relished, but circumstances forced their hand. There were a lot of factors at play, with finances at the root of it, but schooling for the children was also a concern. The bank put pressure on them to sell, then found a couple of German buyers.

JEANINE: The bank actually brokered the sale. I don't know if that was totally ethical. A lot of places out there went under because the real estate market went down and fell apart. In the '70s the interest rates went nuts. We basically were forced to sell. Then I moved out, even though I hated it. I hated leaving that life. I loved that place.

We might have squeaked by if we hadn't had the debt on that big four-wheel-drive tractor. It eventually just couldn't pay for itself. There was never enough work for it. With that kind of work you only do it once. It's not like you come back year after year. Once those meadows are broken, they are good for years and years.

And of course Corky always liked to be off somewhere else doing something. A lot of guys in that country were that way. They just wanted to go. It was a pretty hard scramble for us. We didn't have much extra. Everything we had went into the land. I came out to Rose Lake with Dana and John for two basic reasons. Trying to homeschool John wasn't working. I'm not much of a disciplinarian, and John didn't want to do the work. He would much rather be outside trapping squirrels and a thousand other things. It got to be too much. I was worried because I wanted him to have an education. Also, we needed money. We were down to the bottom of the barrel at that point. I thought I would move to Williams Lake, put the kids in school and get a job. Then maybe when things picked up a bit, I'd come back out. We ended up selling the ranch, and Corky went on working for the guys who bought it.

WORLDS APART: A WORD FROM THE KIDS

JOHN WILLIAMS: I had a hard time with moving from the ranch to Williams Lake. I couldn't relate to anybody out there. I didn't fit in. It was a totally different way of growing up. Living in the bush, I had gone from being young to grown up pretty quick.

Now, at eleven or twelve years old, I couldn't relate to the other kids. We came from such different backgrounds.

DANA WILLIAMS: I was six years old when I moved away with Mother and John. Muskeg had been my entire identity. When we moved away we discovered that we were poor and that we were bushrats. Living on the ranch we never felt like we lacked for anything. Dad worked John like he was a man, and he was a man from the time he was a tiny boy. I don't ever remember him not being a grownup. He's five years older than me, but when we had downtime, oh my God, we played and played and played. We spent all our time outside. We spent time with the Elkins kids next door. We were very close to their whole family. There were always kids there too. It was a great way to grow up.

We moved to Rose Lake in the late fall of 1979, halfway through the first semester, and I remember being floored going into the classroom at the 150 Mile House School. The teacher let me skip the part of grade one I had missed. Because of our ability to read so well, John and I were probably a grade above our peers. They offered to let me go into the second grade, but my mother figured I'd be better off staying with my age group.

Leaving the ranch and moving to Rose Lake sucked for me, but it was really bad for John. When we were little kids we thought we knew exactly who we were and what we were going to be. We were Corky and Jeanine's kids—we were the son and daughter of ranchers. Someday our parents were going to get old, and then we would be ranchers. That's the direction our lives were taking us, and that's what

JOHN AT KLEENA KLEENE.

we wanted it to be. For my brother, this was especially true. To have that taken away from us meant nothing was ever the same again.

My brother spent his whole life trying to get back to that. I had to put it behind me because it was too painful.

Living at Muskeg, we didn't think anything was wrong with us. We didn't think we were poor, and we didn't think we were weird. We were self-sufficient. If we needed meat, Fred would go and pick off a moose. We had a roof over our heads. I didn't think anything was strange about us until we moved out to civilization. Then people were saying, "Oh, look at those kids from the bush." It was rough. We never fit in.

CORKY: I was a hell of a lot better off working for the Germans who bought the ranch than I was trying to make it on my own. I got a paycheque and was making $25,000 a year, plus all my gas and food were included. I had it pretty good for Anahim Lake.

The German owners liked me because I showed them where to go and what to do. We needed that same kind of know-how support when we first came to the country. We had Bob Cohen, Big Fred and Dick Sulin showing us how things were done, and I offered that for the new owners, who didn't have a clue how to function out there.

JOHN POSING WITH JEANINE.

JEANINE: That was the most money we'd seen in a long time. It paid off our debts and Corky would come in every two weeks. The kids and I found a place to live on Branch Drive on Rose Lake and we stayed there a couple of years, then we moved to Alpress Road on the other side of Rose Lake. We still lived in our house at Muskeg after we sold the ranch to the Germans. I came out to Muskeg for two years or so after that, but I could only come in the summer when the kids weren't in school. By that time I had a job and was getting more hours, so it was harder and harder for me to come out.

I worked in the 150 Mile House post office and store. Gary and Dodie Marshall, who owned the store, were just wonderful people. Dodie taught me how to keep books for a small business. After that I did payroll and all the bookkeeping.

In some ways the new owners of the ranch didn't have a great reputation, but Corky deflected that local dislike. He provided the bridge between the new owners and the local people.

CORKY: When the Germans bought the ranch, they got Mike Yates to build them a log house down in the trees as a guest cabin, to give them their own place to stay whenever they were visiting the ranch. I kept

JOHN, DANA AND FRIENDS AT THE ROSE LAKE CABIN.

living in the main ranch house, though Jeanine had moved out with most of our stuff, and Fred and Daisy still lived in the cabin we had built for them. The new owners kept Fred working too, and I was able to hire him on a contract basis.

Things at Muskeg weren't the same once the new owners took over. Some of their ideas were definitely different from what the people at Anahim Lake were used to. For instance, they got a carpenter to install a bidet for them in the new guest cabin, and put a composting toilet in the main house. This caused a curious reaction from Fred.

When I drove out to Williams Lake to pick up the bidet and other supplies for the ranch, Fred stayed at Muskeg to entertain some guests the new owners brought out from Germany to see the property. The new owners were involved in the tourism industry and were always bringing in new investors to look over the ranch. Sometimes a helicopter would land in the meadow and a whole group would arrive.

Fred took them out and showed them around, and they really liked what they saw, but there was a language barrier. The visitors couldn't speak a word of English and Fred didn't speak any German. When

DANA AT THE BRANCH DRIVE CABIN AT ROSE LAKE.

they wanted to come into my house to use the new composting toilet, Fred tried to explain to them that it hadn't been hooked up properly yet and wasn't vented. There was nothing he could say to convince them otherwise, and they came in and did their business. Soon the whole house began to reek with a foul odour.

I never really got onto the indoor composting toilet system, and never used it myself. The composter was powered by propane, and while we had a propane fridge and propane lights, I was still a little leery of having one more propane device in the house, especially after Jeanine's terrifying experience with the propane tank explosion. I don't think we ever got it running properly.

ICE FLOE BEEF DRIVE

CORKY: If you've got to move the cattle during the springtime flood, you want to do it quickly before the rivers get too high. The new owners of Muskeg wanted to run a full-fledged cattle operation so they bought a bunch of cows and we fed them part of the winter at Louie

Squinas's place at Abuntlet Lake. We cut the hay at Louie's place for half the hay crop, then we bought Louie's half of the crop from him, which gave him some income and gave us lots of hay for the cows.

Once we had fed the hay out at Louie's place, we had to take the cattle fifteen miles down the road to Muskeg to finish wintering them before the water got too high. We had lots of hay at Muskeg, but first we had to get the cattle to the other side of the Dean River at Louie's crossing to bring them home.

It was late March and there was just a short section of open water in the Dean where we could ford the cattle. The rest of the river was still clogged thick with ice but it was starting to break up. They always tell you to take cattle into a river at an angle, so that's what we did. We got some of them across but the water was so damn cold that most of the herd didn't want to go. A bunch of the calves had been born at Louie's place and they were still quite young, and they resisted going into that water at all. I don't blame them.

We didn't try pushing the whole outfit into the river all at once. Instead, we bunched them in. When the first bunch got halfway across, we'd start pushing a few more into the river. Several people including Mack Cahoose and his wife, Madeline, and Lenny Leon and his brothers were on horseback. Bob Cohen, Big Fred, Daisy and I were there on foot. Bob Cohen's young son, Patrick Cohen, waited on the far side of the crossing with a long stick to keep the cows from going down that side of the river. He was maybe five or six years old, but he knew what to do. His job was to haze them onto the road to Muskeg, because below the crossing it was heavy duty willows, and it was impossible to take a horse in there to chase the cattle.

Of course, watching over the whole operation was Louie Squinas himself. Bob Cohen said to him, "Pretty big wawa, hey Louie!"

His only response was to holler, "Crazy whiteman!"

To coax more cows into the river, Fred assembled a little box on the back of the tractor and took three or four calves at a time over the ford. He put them down in plain sight on the far bank as an incentive for the cows. We got some cows to cross that way.

Then the cows started milling in the river. Some of them wanted to turn back because their calves hadn't crossed yet. We expected this, so Fred made several more trips with the tractor packing the calves. He

LOUIE SQUINAS.

could carry four at a time if they were stacked in the box right. So more of the mamas went across and we had pretty good luck. But there was a large bunch still resisting the cold water.

Bob Cohen and I had long sticks and kept pushing more animals out into the river, and once they were there we kept them in the water and wouldn't let them come back up the bank. The trick was to keep them as calm as possible and not get too wild with them, or they'd want to come back on you. Once they were across the river we knew they would take the road home to Muskeg.

We had the last part of the cattle in a shallow part of the river, where it wasn't too far across, when I heard Mack Cahoose call out, "Look what come!" All of a sudden a big section of ice shelf that was jamming the river upstream broke off and started floating straight for the cattle. When it let loose, big chunks of ice were thrust into the air, and the whole thing broke loose in one chunk. Damned if the lead cow didn't look up, see the ice coming and step up on the moving ice floe. Then the other cattle followed suit and they fought like hell to get on that iceberg with that lead cow.

The whole outfit started floating down the river. The worst thing was that there was a fourteen-foot waterfall about three-quarters of a mile farther downstream. The bottom of the waterfall was deadly, with huge rocks waiting to grind them up. To complicate it further, the river leading to the falls was fast and the banks were quite steep.

As the whole works of the cattle started to sail right on past us, Fred had his hand over his face. He said, "I can't watch. It looks like they are sailing to Hawaii."

I said to Bob Cohen, "Did you ever see anything like that! Them sons of bitches fought to get on top of that ice, and the calves did too." As more cows got on the iceberg, the ice started to submerge, then to break up, and the cows fell in. So you've got cows floating over here and floating over there. The whole river was full of cattle, and there wasn't a thing we could do but look. The last we saw of them, the whole outfit disappeared around a bend and was heading for the falls. I remember Bob Cohen saying, "Just when you think you've seen it all, by God you haven't."

There wasn't any use chasing them because once they got out of the river, they would start running through the timber. It was a long ways down the road to Muskeg, but the cattle knew the way. By a stroke of luck they avoided going over the waterfall but they were scattered like mad, on the fight and on the run when they got out of the river. Miraculously, every one of those suicidal cows made it home alive.

KANGAROO COURT

SAGE: Before things got too civilized and formal around Anahim Lake with the construction of the new, modular provincial courthouse, all legal proceedings were held in the community hall down at the stampede grounds. One day a dispute between two ranchers over a range permit had to be settled, and Corky, Bob Cohen, Francie Wilmeth and Roy Graham figured the hearing was shaping up to be a sham. They weren't exactly fans of government anyway, so it didn't take too much convincing to get Corky to make a statement about the event in the best way he knew. Theatrically.

JEANINE: What I remember about that was that Forestry had put in some new rules about grazing, and the ranchers were always fighting about grazing rights. In this particular case the community felt that one of the ranchers was getting screwed by Forestry, because they were going to take his grazing rights away or force him to drift fence it.

SAGE: Bob Cohen recalls that day of the range permit dispute. "It sounded like a kangaroo court to us," he says, "so we ordered Corky a kangaroo suit from a costume shop in Vancouver. Then we dressed him up in it and took him up to the community hall where everything was going on. Corky, being a rancher, was expected to be there. We wired his hands so he could wiggle his ears and he had a bottle of whisky in the joey pouch. Before we took him into town we had him out in the yard at Muskeg practising his kangaroo hop, and Fred had just come home after being away on a toot. He took one look at the kangaroo in the yard and never said a word. He just went into the house and started drinking coffee.

"When we brought Corky to the community hall, he tried to get me to come inside with him. I said no way. Court is one place I don't want to be. Besides, there was no way I wanted to get involved with it because it was a personal dispute between two ranchers and I was breaking horses for both of them. So I had to keep out of the damn thing.

"They had a roll call for the ranchers, and when they called out Corky's name, he stood up in his kangaroo suit. One of the lawyers said, 'I've heard of kangaroo courts before, but this is the first time I've ever seen a kangaroo.' Not too many people would have had nerve enough to go to court dressed like that."

CORKY: That costume was awkward as hell with its big, long feet, and I had to learn how to move with it. You had to kind of hop up in the air to move forward. The tail was a-dragging, so we had to wire it up. I had a hell of a time getting into the hall. That kangaroo's tail kept jamming in the door. Bob and Francie had to hold the door open for me and I finally I got through. Then I was on my own. Bob and Francie and Roy waited outside by the corrals and kept an eye on the proceedings through the fence rails. Every once in a while I'd go outside and have a big drink and wave at them.

The lawyer for one of the ranchers had a hotshot reputation. He did a lot of pacing and kept his head down while he talked to the judge. He didn't look up or take notice of me for the longest time. I was in the back row, and he kept pacing back and forth and speaking to the rancher on the witness stand. "Well Mr. So-and-so, how long have you had this grazing lease?"

Finally he looked up and saw me, and his jaw fell open. I had just opened the joey pouch and was taking a drink, and the bizarre nature of the situation staggered him. That was our whole intent. He wavered as he spoke to the judge. "What was that last testimony, Your Honour?" he asked. From that point on he refused to look at me.

ON CORKY TEARING THE SHIRT OFF THE FORESTRY GUY

SAGE: Even though Corky no longer owned Muskeg Ranch, it did nothing to ease his frustrations with Forestry. There was one official at the Forestry office in Alexis Creek who was particularly officious and seemed to have it in for Corky, and they had several heated confrontations.

CORKY: This guy kept screwing with my cattle permit, always trying to give it to somebody else, so we had a squabble about it. He wouldn't increase my permit even though I had done all the work I was required to do. I'd put up fencing, complied with all the regulations, and done everything they asked me to, then this meathead came up with a bunch more excuses not to give me my permit.

I drove all the way to Alexis Creek see him, which was 150 miles and several hours away from Muskeg. I told him I figured he was the most incompetent son of a bitch I'd ever been around. "You don't know your ass from a hole in the ground. You're keeping us from developing our places with all these damn laws you've got." Then he got lippy with me and I finally blew my top. I grabbed him by the shirt, told him he was nothing but a government parasite and slapped him around the head a couple of times. We bounced off a couple filing cabinets in the Forestry office, and his shirt got torn. A secretary walking by with a load of paperwork dropped everything she was carrying and papers were flying everywhere. My daughter, Dana, was with me, and she saw it all. I remember looking back as we stepped out of the Forestry office. I can still see that guy standing there in his shredded shirt.

DANA WILLIAMS: I was about nine years old when that happened. I wasn't living at Muskeg anymore. It was during the summertime and I had gone home during holidays to visit with my dad, and he was all haired up about a range permit Forestry didn't grant him.

We drove to Alexis Creek and went into the Forestry office. I don't remember exactly what was said, but finally Dad hauled off and said, "You goddamn son of a bitch, I'll show you," and he punched him right in the face. He hit him hard then grabbed him by his shirt and shook him all around and threw him into the filing cabinets. Shit flew everywhere. I

CORKY HAS ALWAYS BEEN THE ARGUMENTATIVE TYPE.

remember thinking, "Oh crap, this is bad!" Dad grabbed me and said, "Come on, we got to get out of here." We roared down the street in Dad's pickup heading out of town and then he remembered we had to get gas because we were about on empty. So we turned around and raced back to the filling station.

The attendant said the Mounties had just called. "They're after you," he said. I remember wondering where we could hide. Dad said, "Get down. Get down on the floor!" While we were fueling up the cop came along. I was surprised—the officer didn't seem too upset about what had happened and he let us go. After we got the fuel we headed back home.

I can't tell you how many times in my life I watched my father punch somebody out. He punched out a guy in line at the bank because the guy called him a half-pint. He's been in so many fights in his lifetime, I literally can't count them. He'd fight with anybody; he wasn't afraid. I can still remember the colour of the Forestry guy's shirt. The memory is clear as day, because it was so vivid and crazy.

That's how it was with my dad when I was a little kid. We would leave the house and you never knew what crazy shit was going to go down. We accepted that about him. One time we were at Rose Lake and Dad said, "Hey, let's put on wigs and go to the drugstore and not even tell people why we have them on." So we went to Spencer Dickie's drugstore with wigs on. We went to the grocery store when I was seven years old and he gave me a blank cheque and told me to get what I thought we needed. I went around through the aisles. Usually we needed canned corn, so I got that. Then when I got to the checkout I handed the blank cheque to the cashier and told her my daddy said to get her to fill it out. He would do crazy stuff. He didn't care what people thought of him.

CORKY: I was topping up my tank at the gas station in Alexis Creek when the cop came along. Forestry had obviously called him. When I told him the story about how impossible this guy was to work with, and how difficult he was making it for ranchers, the cop smiled. It turns out that guy wasn't very popular with other people either. He didn't last too long at Alexis Creek.

I got a letter from the cops a week or so later saying they weren't going to press any charges against me, but asked if I could find it in my heart to pay for his shirt. I'm not sure if I ever did get around to buying him a new shirt.

LESTER AND PAN: THE END OF AN ERA

SAGE: Just over two years after Jeanine and the kids moved to town, Lester Dorsey passed away at the relatively young age of seventy-eight. No one had impacted the lives of Jeanine, Corky and their kids more than Lester during the decade they knew him.

JEANINE: Lester Dorsey used to wear the most incredible outfits because he never paid any attention to what he had on. He'd pick up whatever was handy. I would have not been surprised to see him wearing a tin pot on his head. He didn't care. He just did whatever. He was awful about taking your things and wearing them. He'd take your socks, your gumboots or your gloves. He was death on gloves. He liked those woolly scarves, too, and you really had to watch him

LESTER DORSEY ESSENTIALLY OCCUPIED THE WHOLE LAND.

when he was ready to go, because they'd disappear and you never got them back. I got to where I'd hang around while he was dressing himself to leave.

Lester essentially occupied the whole land. He'd been everywhere, he knew the whole country, and nobody but Cohen knew it better. He was a good hunter and a good rider, and he loved his horses.

He was something else, and he loved gossip. He'd gossip worse than any woman, and he was curious as a cat, always wondering what was going on.

CORKY: Lester was a character. Our place at Muskeg was a good place for him to stay all night because it wasn't very far to the Clark Place. Only four miles down the river. Only fifteen minutes if he was travelling on snow machine.

Lester would keep you up all night telling stories. The hunters who paid big money to come out on a ten-day bear hunt didn't care if they got anything. They came up to get drunk with Lester Dorsey. They loved that.

MICKEY DORSEY. LOSING HER BROKE
LESTER'S HEART.

DANA: At times Lester was a grouchy old codger, but he and I got along like peas and carrots. Just before he died, he and Dad and I went on a trip together. We went all over BC and drove into Washington State. At the time I didn't know what the purpose of the trip was. Later I found out it was for Lester's chelation treatment, and he died shortly after. We had so much fun on that trip. I loved him. He was one of my favourite people in the world.

CORKY: In the spring of 1982 I got a letter from Pan Phillips dated April 20. It read:

Hi Corky, how did you winter? It is a winter to remember. I understand your land clearing machine is over at Peter Alexis's place. I would like to have you rotovate some land for me, and I think Wes Carter wants some done. William Cassam wants some done, which I believe is Indian Department. I also talked to the Home Ranch owners and they sounded interested, they might have as much as 640 acres. How do you charge, by the hour or the acre, and how much? I hear Lester Dorsey is in the hospital again or is he out by now? Hope to hear from you or maybe see you. Sincerely, Pan Phillips.

The letter was typed and he signed it, "Pan."

SAGE: A bit over a month later, on May 24, 1982, Lester's wife, Mickey, died when the plane she was riding in flipped over while taxiing into the bay at Port Hardy. Mickey was dying of cancer and was on her way to receive treatment in Vancouver when the accident happened. The combination of Lester's poor health and the shock of losing Mickey so unexpectedly was tough on him. A month later Lester died of heart failure in the Williams Lake hospital.

Less than a year later, Pan Phillips also died. He keeled over on May 28, 1983, while listening to the messages being broadcast on Cariboo Radio. In those days, Pan and many other people living in the backcountry, from Nemiah Valley to the Blackwater River and from Barkerville to Forest Grove, stopped whatever they were doing twice a day to listen attentively to message time. Pan's lodge was full of fire-fighters at the time, battling a forest fire east of his place. His daughter, Diana Phillips, writes in her book *Beyond the Home Ranch*, "Dad would have been in his glory, not only having an audience to entertain, but his cabins would all be rented out, bringing in an income."

The passing of three legendary characters whose lives carved such a mark around Anahim Lake was the end of the final chapter of a unique time. An era had quietly yet dramatically come to a close.

CORKY'S LIFE-CHANGING OUTHOUSE DEBACLE

SAGE: Anahim Lake Stampede is the one social event of the year when the world comes to the community's door. It's always the second week-end in July, following the Canada Day long weekend, and the local people pitch in to make the Stampede an event to be remembered. There's a gymkhana, a bullarama, two days of rodeo and four nights of dancing.

In 1985, volunteers had done the usual preparations for the Stampede, making sure the stock chutes, arena surface, holding corrals, fences, concessions and dance hall were up to snuff to handle the big influx of people and animals expected. One of the improvements that year was replacing the old outhouses in the stampede grounds with galvanized steel culverts placed upright in the ground. Then the original wooden outhouse structures were set back over top.

Because booze inundated every part of life around Anahim Lake, Corky sometimes took a hiatus from drinking for his own health and well-being. But it wasn't easy. That's how it was in 1985 when the Stampede rolled around. Corky says he hadn't had a drink for several months, and he wanted to keep it that way. He still wanted to visit a few old friends who always showed up at the Stampede, so he camped at Bob Cohen's place in Anahim Lake where several of his friends were staying. Instead of heading to the rodeo with everyone else, where he knew the liquor would be flowing, he chose to spend the day back at camp and take it easy.

CORKY: When I tell people what happened, they want me to tell it to them over and over again, because they never heard of anything like that.

JEANINE: Everybody said, "Well, nobody but Corky." If it was going to happen to anybody it would be him. Somebody had moved the outhouse building at the stampede grounds off the hole, and Corky never noticed. Whether it was a prank or carelessness, we were never certain, but when Corky went to use it, he had a serious fall that affected the rest of his life.

CORKY: Contrary to what some people think, this happened when I was sober. It was Sunday, the last day of the rodeo, and Maurice Tuck's wife, Diane, and I stayed up at Bob Cohen's place all day where we were all camped out, while everybody else went to the rodeo. I waited until about seven or eight o'clock that evening, then told Diane I was going to walk down to the stampede grounds to see what's going on. I knew the rodeo would be over and the crowd would have thinned out by then.

Maurice Tuck and Bob Cohen had been saddling and unsaddling the stock all day at the stampede. This was something they had done every year for the past twenty-five or thirty years. I wasn't too interested in going behind the chutes to see those guys because I saw them every day, and I knew they'd try and get me drunk if I showed up there. So I waited until the crowd had thinned out and the cowboys had collected their rodeo money and got the hell out of Dodge.

When I got to the Stampede grounds there were still a few people around the beer garden. I needed to go pee, so I headed over to the outhouse. It was still quite light out so I had no problem seeing where I was going, but when I got to the outhouse it was occupied and I could hear people talking inside. I waited and waited, then when nobody came out, I stepped behind the outfit to take a pee, and down I went.

Somebody had gone and shook the hell out of the outhouse and moved it off the hole, and the top of that eight-foot culvert pipe was half exposed. If I had gone in headfirst I might have drowned in the shit.

There were all these rocks piled up around the top of the pipe and I must have slipped on a rock and took a header. As I started to go down I could see that big cavern of shit waiting to suck me in. I clawed the back of the outhouse to keep from going in the hole, but there was nothing to grab onto, so I made a big effort to miss that son of a bitch. I turned at the last second and somehow missed the hole, but my shoulder landed hard on the edge of the culvert and I knew I was in trouble.

Somebody was standing there and I asked them to run down and get Bob Cohen. Bob is real smart about that kind of stuff. When he got there and saw how badly I was hurt, he said I was likely to be in shock and they had better get me right down to the nurse. Cohen and Duke Sager loaded me into Duke's van and took me to the clinic. The nurse was completely incompetent. She only gave me Aspirin for a broken arm and wouldn't give me any painkillers because she couldn't get hold of a doctor.

I spent the night on Bob's couch, and when you get hit that hard and go into shock the adrenalin will keep the pain away. But if you move just a fraction of an inch, oooh! I didn't know it at the time but I had three shards of bone in my upper arm sticking into the nerve. The next morning Bob's wife, Francie, called down to Dean River Air Services to get Floyd Vaughan to fly me to Williams Lake. Roy Graham drove me down there and Floyd loaded me into his Beaver and we headed to the hospital. We got into a storm about twenty miles from the Fraser River and it threw me around inside the plane like a pinball. I couldn't wear the seatbelt because of my arm, so I got thrown around pretty good. It was the roughest weather I'd ever experienced in a bush plane. Somehow I got hold of the strap and managed to hang on. The plane would drop, then go up, then turn. It beat me up pretty bad.

From Williams Lake they sent me on to Kamloops. The doctor showed me the x-rays and the shoulder blade looked like you'd hit it with an axe. It was broken in a straight break; my arm was just hanging there. It crushed the nerve and it had to regrow.

JEANINE: Corky was casted up for quite a while, maybe two months. He broke his scapula right in two, but more seriously, his upper arm was broken in a twisted spiral near the shoulder, and he suffered a lot of nerve damage, so it took quite a while to heal. It was a year before he was able to use his arm again. It took a while for it to sink in how bad the injury was, especially with the nerve damage. After several months, Corky realized he was not going to be able to do the kind of labour he had been doing that's required on a ranch. His arm was never going to be quite good enough to do that kind of work, and he would have to quit our old ranch at Muskeg.

CORKY: I got hold of the Germans and told them I'd been hurt. They were really nice about it and wanted to know how serious it was. I told them they better get somebody to run the ranch because I was unable to. I could barely move for several months, then I had to go to a therapist in Kamloops every six weeks to make sure everything was healing right. The nerve grew back but it took a long, long time.

JEANINE: We filed a lawsuit against the Anahim Lake Stampede for damages. It should have been a simple case covered by the liability insurance of the community association, but the insurance company fought it all the way. It took several years and we had to travel to Vancouver for the court hearings. We finally got it done but we didn't get anything near what we thought we would. Our lawyer said the judge gave incorrect instructions to the jury, so they only awarded us a quarter of the damages. Our lawyer wanted us to appeal because he said he absolutely knew we could win. But we didn't have any more money left. It would have cost four thousand dollars just for the transcript from the trial to proceed to Supreme Court. Our lawyer wasn't a big firm. He was just a single guy working for himself, so he couldn't afford to work for nothing either. It was a heartbreaker.

GLEN KING KILLED IN PLANE CRASH

CORKY: A few days before I got hurt at the Stampede, Ollie Moody came down to Muskeg to help me plough up some land. A forest fire got started by a lightning strike the night before, high up in the Ilgatchuz Mountains right above the ranch. We were standing in my front yard watching it and the smoke was boiling everywhere, so thick so you couldn't see through it. Then we saw a little plane flying along the edge of the mountains heading toward the fire. It flew into the smoke at the top of Saddle Mountain but we never saw it fly out. I turned to Ollie and said I couldn't believe what we just saw. I figured that plane must have crashed. We soon found out it had. Glen King, the guy I sold the Lessard Lake property to, was killed, but his daughter, Annette, and the guy flying the plane survived. We later found out the pilot was Bob Hutchison, the owner of the plane.

OLLIE MOODY: Corky said, I think that son of a bitch crashed. We listened. You could hear the roar of that fire from Corky's place, but we couldn't hear or see the plane. We stood right there and watched it. It was a clear day too. They must have flown in over the fire and lost their air. If you got no air, the plane won't fly. The pilot was in tough shape for a long time.

CORKY: The smoke was covering quite a big area but we had a real good view of where that plane went in. I had the CB radio turned on in case there was any news about the fire, and I heard Lora Vaughan calling around asking if anybody had any news about a missing plane. I told her what we had just seen and it wasn't very long until that mountain was crawling with people. They parachuted people in there because they knew it was a crash. For a verified crash, they throw the whole ball of wax in there.

OLLIE: They helicoptered paramedics in there, bang, bang, bang, right now. We saw it from Corky's yard. It was like watching a movie. It was right there. The hard part was finding out it was our neighbour, Glen King, who I knew from taking a beef drive around the mountains from the Blackwater a few years earlier.

They flew right into that mountain, kapoof! It's a miracle Annette only got a broken arm and a few cracked ribs. She was really lucky. She was all strapped in, but it ripped the seat right loose from her. Annette saved the pilot's life and she was only nineteen years old. I heard later the pilot recovered but killed himself in another plane crash three years later.

CORKY: A few days after the plane went down, I was in the hospital recovering from my outhouse accident at the Anahim Lake Stampede, and a Mountie came down to talk to me about Glen King's crash. He wanted to know what I saw. I told him I'd been listening to the mes-sages on the CB radio to see if there was any news about the forest fire. I told him how we saw a little plane flying along the edge of the mountains and disappear in the smoke and never saw it come out, and how the next day Forestry got involved and there were gobs of guys fighting the fire.

CORKY BACK ONSTAGE

EXPO 86 AND BECOMING A STORYTELLING ENTERTAINER

CORKY: Ian Tyson came to Williams Lake in 1985 to hold auditions for his *Cowboyography* show at Expo 86 in Vancouver and he put a notice in the paper that they were looking for local talent. They wanted the whole province represented and he was going from town to town up and down the province looking for storytellers and poets. I met Ian at the Chilcotin Inn in Williams Lake where he held auditions for two days. Somebody from Riske Creek who knew Ian heard of me doing some poetry somewhere, and told him about me. Ian chose me and Katie Kidwell from 100 Mile House. She was a real neat lady from an old ranching family, and she and I got to go to the World's Fair. They put us up and paid for it all, gave us money and food, and we took it. We did twenty-one performances at the World's Fair in ten days, and we filled up the audience too.

JEANINE: Dana, John and I went down as a family and stayed with friends on Bowen Island. We took the water taxi to Vancouver and that was some fun. I'd never been on a small boat on the ocean before. We went all around the Fair, and did this and that, and would come by to where Corky was performing and say hi.

CORKY: They always said Ian Tyson was hard to get along with, but he was really nice to me. It was musicians who weren't doing what he wanted who complained the most. He wanted you to do it good, and if you weren't doing it good, then do it again until you get it right. He was a perfectionist. One night one of the performers in the show went out on the town, and he wanted me to go with him. Thank God

CORKY IN HIS BUSHMAN COSTUME, PERFORMING AT EXPO.

I didn't go. He wanted me to go out honky-tonking with him, and I knew I would only get drunk if I did that. I didn't drink anything while I was at Expo. Next morning that son of a bitch showed up in the middle of the rehearsal call all hungover, eyeballs hanging out. Ian chewed him up one side and down the other. He said, "I'm holding your pay until you do it right. Don't ever do that again." That stopped it right there. But Ian wasn't cranky as long as you were doing your part.

I hadn't been on the stage for fifteen years when I performed at Expo. After that, people wanted me to come around and do things for them. They didn't know I even existed until I performed with Ian Tyson, and Ian was a big help in promoting my act after that too. My next big engagement was in Las Vegas, where Johnny Drift had me come to entertain the guide outfitters.

JEANINE: John went to Vancouver right after he graduated from high school that year and got a job working in construction. A couple months later Corky went to Vancouver and got an agent, and he and John shared a grubby little apartment on Robson Street before that part of town got its facelift. There were lots of working ladies in the neighbourhood.

John started on ground zero with the construction trade, and did very well. He worked

TOP: CORKY PERFORMING IN WILLIAMS LAKE FOR THE GUIDE OUTFITTERS. BELOW: CORKY ON *The Beachcombers* SET

178 — CORKY WILLIAMS

independently and knew about all kinds of machinery and power tools. He was a cut above what they expected from a kid fresh out of high school. The guys he worked for wanted him to stay so they could make him into a foreman. They were hinting there might be something for him beyond that, but John did not want to live in the city. He wanted to come back to the north country.

THE COLOURFUL CHARACTERS OF ROBSON STREET

CORKY: A string of prostitutes lived right next to the place where John and I were staying. We were up on the fourth floor, and all the hookers and pimps lived down on the first floor. In the middle of the night sometimes people would be screeching out there below my window. These buildings aren't very far apart, so the sound goes right up and in your window. We had to keep the windows open because it was too damn hot if you didn't. This one time there were people yelling down there, so I hollered at them to quiet down. When they ignored me I started throwing things to get their attention. I dug the garbage out and I fired a couple of tin cans, but I missed. Finally I got a tin of Dr. Scholl's foot powder and fired that down. It hit the cement above and blew the lid off, and the thing exploded with white powder. That quietened them down. That foot powder bomb worked like hell.

I got an agent while I was in Vancouver and she was starting to get me work. I started getting beer commercials and parts on television shows like *The Beachcombers* and *Bordertown*. They built a little western town out in the boondocks near Albion in the Maple Ridge area, where *Bordertown* was filmed. They shot everything right there.

I was still recovering from my accident the year before at the Anahim Lake Stampede, and I was able to get the physiotherapy treatments I needed for rehab in Vancouver. I saw my agent nearly every day and I went to therapy sessions three times a week. The nerve in my arm was starting to regrow but it was a long, slow, painful process. I went to over one hundred therapy sessions over a couple of years. When your hand is all twisted up and the nerve finally grows, your fingers are all clenched. The therapist moves your fingers just a little bit at a time. I would do exercises in a swimming pool and that helped a lot.

LUTHER "Corky" WILLIAMS

IN VANCOUVER CORKY GOT AN AGENT AND HAD HEADSHOT PHOTOS TAKEN.

CORKY MOVES TO ALBION

JEANINE: At the end of the summer John decided he'd had enough of city life, and he moved back home to the Cariboo. Corky couldn't afford the rent on Robson Street by himself, so he moved to Albion—he

ABOVE: TWO OF CORKY'S ACTING BUDDIES. DAVIE LONGWORTH IS AT LEFT.
BELOW: CORKY ON A MOVIE SET IN ALBION.

was good friends with some of the guys there working the stock for *Bordertown*. Another cowboy poet, David Longworth, was there, and they all lived together in a shared house.

CORKY: I was in Albion for four or five years. There was this guy, Les Harris, from the Snowy River country of New South Wales, Australia. He was a working cowboy at the Douglas Lake Ranch when they filmed the Billy Miner movie up there. He was working on the ranch, and whoever was casting for the film saw him and went nuts over how he looked. He immediately wanted to use him in the Billy Miner film, so they did. I got acquainted with Les because he and I were going out for the same kinds of parts. He had that look. That's how the film industry works. I'd run into him all the time at auditions. He had a different kind of cowboy look, with big mutton chop sideburns. Once the Billy Miner filming was over, he went over and talked to an agent. He and I became good friends. He was married to a Native lady from Merritt.

DANA MOVES TO TEXAS

JEANINE: After Dana graduated from high school in 1991 she immediately packed her bags and was gone. She moved out of Williams Lake, shook the dust off her feet and went down to live with her uncle, (Jimbo) Jaston Williams, Corky's brother, and his partner in a beautiful apartment in an old historic building near downtown Austin.

DANA: I came to Austin right after I graduated. I had a serious boyfriend, lots of really good friends, and I had all kinds of stuff going on in Williams Lake when Jimbo called me and said, "You want to come live with me for a year before you start college? I'll pay for it. You come stay with me for a year." And I said, "Yeeaah!" I went in the kitchen and told my mom I was going to live with Jimbo. She looked at me, and the way she described it to me years later was that I had that look on my face where I wasn't asking her permission, I was stating how it was going to be. She said, "Okay."

It wasn't long after Jimbo's phone call that I left. I arrived in Austin on August 18, and my uncle was on the road travelling with his show, so for the first three months I was in his apartment by myself. It was fantastic. An open bar, this knockout apartment in downtown Austin. It was like getting let out of jail. It was wonderful.

CORKY MEETS THE DAUGHTER HE NEVER KNEW HE HAD

JEANINE: Corky was living in Albion when he found out he had another daughter. When he was in high school in Van Horn, Texas, he was going out with this girl, and then she moved away quite suddenly. Corky went on with his life and didn't think too much about it. Thirty-six years later, while he was working on a set in Albion, he got this call. A woman asked, "Are you Corky Williams?" He said, "Yes." Then she said, "Well, I'm your daughter, Raylene."

CORKY: I was working at Albion Ranch when I got the call from Raylene. I thought it was just the guys pulling a joke on me at first, but I realized it must be true because she knew details they never could have known. I was standing on the stairs with the phone on an extension cord, and I damn near fell over. I couldn't believe it.

It's a funny thing, how she found me. She had a friend she went to school with who had moved over to East Texas, and she went into the post office in Van Horn and asked how they could find me. Well, there just happened to be a woman standing there who went to school with me. She said, "I know him." And that's how it happened.

I called Jeanine and she called Dana, who was really excited because she had always wanted a sister. Eventually I flew out to Phoenix and they were waiting on me. By that time Raylene had met a big passel of my family. My dad and Adele had a family reunion, and my brother and sister got to meet her. When I got to Phoenix, not only did I have a daughter, I had three grandkids as well.

Discovering our family worked out nicely for Raylene because by that time her adoptive father had passed away and her adoptive mother was sliding off into dementia. So she lost that family and found our family.

JEANINE: You take one look at Raylene and you know. She looks almost like Corky's sister. A lovely, lovely woman. Dana was the first one in the family to see Raylene, who lives in New Mexico.

RAYLENE, SECOND FROM RIGHT, AT HER DAUGHTER'S WEDDING.

It turns out when Corky's high school girlfriend realized she was pregnant, she left Van Horn, because in the mid-1950s, what do you do? She went to stay with an aunt in New Mexico, had the baby, then handed the baby over to the couple who adopted her.

There's a long trail of people that led Raylene to Corky. This one telling that one, asking this one, and it somehow got back to Van Horn and somebody said, I know Corky, he moved to Canada. It was amazing that she was able to track Corky down to this little shantytown movie set they had built for *Bordertown*.

Dana goes over to see her at least once a year, and of course Raylene comes to the weddings and various family get-togethers. It's so wonderful. I thought Corky's parents might be a little leery of all this, but they opened their arms up and said, "Come on in." They welcomed her like a long-lost loved one. She went to meet Corky's dad and his second wife, Adele, when they had a family gathering out there. Raylene said it was the most wonderful day of her life. It was like she was always there, a part of the family.

DANA AND RAYLENE.

DANA: I was excited beyond belief to hear about Raylene. I remember when my mother called and told me the news. I was with my Uncle Jimbo in Austin. We were taking down the Christmas tree and the phone rang. Jimmy answered it and I heard, "What the hell! Holy shit!" I remember thinking, "What has my dad done now?"

Jimmy gets off the phone and says, "You have a sister and she's thirty-six."

I said, "Dad never told me I had a sister."

"He didn't know either until just now," Jimmy said.

I had to call my dad right away. When I talked to him he was still on the ground. He said, "I'm here on the carpet with the phone and I don't know what to say."

I was eager to meet Raylene. We arranged to meet in Arizona, and it was great. I remember landing in Phoenix and getting off the plane. I saw her across the airport and even if I didn't have a picture of her, I would have known she was my sister. We look very similar. To me she looked like an older, brown-skinned, brown-haired version of me. She's very "Williams"—she is short and stocky, and she has the classic Williams face and Williams smile. I ran across the airport and hugged her and it was just as if I had known her all my life.

CORKY AND JEANINE PART COMPANY

JEANINE: Corky and I split up in 1990. We just reached a point where it was time to let go. When we split up I moved to town and found a place to live in Williams Lake and Corky stayed on at Rose Lake. Then he moved down to Albion full-time.

Corky had already gone to Texas and was living in San Antonio when they closed the post office at 150 Mile and moved it into the brand new 150 Mile Centre just down the highway. Without the post office there wasn't enough work for me to work full-time at Marshall's store, so Dodie laid me off so I could go on unemployment insurance.

They had a special program for women entering or re-entering the workforce who wanted to upgrade. So I joined in that. I took an accelerated course where I got an associate bookkeeping diploma in one year instead of two. Then I went to work for Jackpine Lumber, a local lumber manufacturing plant in Williams Lake, but I wasn't happy there. John was off working up north, and Dana was in Texas. There was nothing holding me in Williams Lake at that point, so in 1993 I decided to move back to Texas too. I headed to Austin and Dana and I lived together for a year, then I moved into a place on my own.

I got a job working for the state government in Austin, where I worked two years as a temp. Then I got hired on full-time, and had to work for eleven years to qualify for my pension. I came back to Williams Lake in 2007, and by then Corky was already living in Canada too.

CORKY: Booze was a contributing factor in our splitting up. Booze can do that to you. Another problem was that I was gone so much, trying to make a living. Being gone all the time created problems, because I wasn't there to do my part. Our splitting up happened, and it's unfortunate as hell, but you have to live with it.

NUDIST CAMP

CORKY: One winter when I was at Albion I went down to visit my brother, Jaston, in San Diego, because he was putting on a play in a fancy theatre there. I don't like to fly, so I took the bus, and I had a really good time sitting at the back with a bunch of Canadian soldiers who were heading south for some winter sunshine. We drank whisky all the way to California.

I met up with Jaston in San Diego and went to his play in the famous La Jolla Playhouse. That's where I ran into Ardyth Clements, an actress

CAST PHOTO OF CORKY (SECOND FROM THE LEFT) AND ARDYTH CLEMENTS (SECOND FROM THE LEFT IN THE BACK ROW) IN A PRODUCTION OF TOBAC-CO ROAD BEFORE CORKY LEFT CALIFORNIA. ARDYTH LATER INVITED CORKY TO A SAN DIEGO NUDIST COLONY.

I had known in Los Angeles years before and had been on stage with a couple of times before I moved to the Chilcotin. She had been married four or five times since then, and was married to this retired profes-sional wrestler who went by the ring name of Doctor Death. He was a giant of a man with muscles in his toenails. He and Ardyth had moved into this nudist colony outside of San Diego.

They invited me out there. At first I didn't want to go, but she got me half-drunk and talked me into it. I thought I better go and see what's going on.

I got a bottle of Rebel Yell cocoa liqueur. It was strong but you mix it with milk, and I had a big blast off that. We drove out in her car and she pulled up to these gates that were about twelve feet high to keep the yahoos from looking through at the nudists. She hit her button and the gate opened to a chain-link fence covered with ivy, and the road followed along beside it. She parked her car and through the ivy I could see flashes of flesh going by. I was watching pretty good, and I asked Ardyth what the hell was going on. She said they were playing tennis.

"You mean to tell me you've got people here playing tennis with no clothes on?"

"Oh yeah, yeah, yeah. Hang on, we'll go around the fence."

So here they were, a bird's-eye view, a couple of eighty-year-olds all naked as hell playing tennis. Another woman without a stitch on except a pair of tennis shoes and a rain hat was standing in her yard raking leaves. Ardyth knew them all. I had to meet all of these people, and of course I'm sipping my Rebel Yell the whole time. Ardyth told me people who are guests at the nudist camp are allowed to remain clothed, but of course you're encouraged to take off your clothes and be natural.

I went up with Ardyth to her apartment and met Doctor Death. He didn't have a neck, just a stump on his shoulders. His chest was like a bull, and his arms and legs were as big around as my body. He was around sixty-five and on a strict fruit diet and didn't eat any junk food. He told me all about the wrestling business and how it works. He was a psychiatrist, and one of the first men to ever lift six hundred pounds. He was also quite musical and sat there buck-naked, playing the piano.

The next day Ardyth gave me a big towel and said we should go down to the pool. I said, "I ain't going." Then after a little more whisky I said I'd do it. I had my plastic jug and this big long red towel probably about six feet long, cut like a poncho. When we got down to the pool I immediately piled into the deep end, towel and all, to hide out. But the plan backfired. All around the pool were these bodies laid out horizontal on sun chairs, the whole outfit bare-ass. When I bobbed up, all I could see was people naked as hell lying in the sun. I was surrounded. And you could see everything.

Ardyth was crazy as hell. She came over to the pool and said, "I want you to meet this preacher, he's from British Columbia." I'd met reverends and their wives before, but I'd never been introduced to any that were naked. There I was at the edge of the pool, and the reverend's wife did a little curtsy and I'm looking up and I could see all the way to Bakersfield.

Ardyth wanted me to stay over because the next day's entertainment was going to be this naked bluegrass band, but I'd seen enough. I told

CORKY PERFORMED IN CITIES ALL OVER THE UNITED STATES. OPPOSITE
TOP: CORKY ONSTAGE IN ARSENIC AND OLD LACE.
OPPOSITE BOTTOM: CORKY RODE OREO, A PUREBRED LONGHORN STEER, IN
A PARADE IN BANDERA, TEXAS.
ABOVE: CORKY AND HIS BROTHER JASTON IN A FUNNY THING HAPPENED
ON THE WAY TO THE FORUM.

her I had to get out of there. My comfort zone was maxed out. The name of the nudist outfit was The Swallows.

CORKY GETS A PART IN THE FOREIGNER AND MOVES BACK TO TEXAS

CORKY: One day Jaston called me at Albion and said he had a part for me in *The Foreigner*, which was the next play he was producing. Jaston and his partner, Joe Sears, owned a professional theatre company in Austin, Texas, called Greater Tuna. They were very successful, and wrote and produced plays on all the great stages across the United States. They started out as just a couple of hippies in the early 1970s. I had been going back and forth between Texas and Albion for a couple of years, so when I went down for the audition and got the part, I decided to move back to Texas. I got an agent in Austin and found a place to live in San Antonio, and our run with *The Foreigner* lasted three or four years.

I was cast as this Ku Klux Klansman, and I was real nasty. We took *The Foreigner* on a national tour and played in all the top theatres. One of the places we played was the Ford Theatre in Washington, DC, where Abraham Lincoln was shot. We played in Washington for a few weeks and I got acquainted with a bunch of the theatre staff. At first they kept their distance whenever they ran into me, but they finally got it that in real life I wasn't as fiendish as the character I portrayed. I must have played the part pretty convincingly because every black person who worked there came to watch my final demise and then clapped like hell with big smiles on their faces. Afterwards I went up and shook their hands. It was a good experience.

WORKING IN THE BUCKHORN SALOON AND MUSEUM

CORKY: One of the places I worked in Texas was the Buckhorn Saloon in downtown San Antonio. It's an old-time Texas saloon that's been there since the middle of the 1800s. It was started by a seventeen-year-old bartender and bellhop by the name of Albert Friedrich, who decided to open his own saloon. It wasn't long before he realized that many folks travelling the ol' dusty trail didn't have much money, so he started accepting horns

HEADSHOTS FROM CORKY'S ACTING PORTFOLIO.

and antlers in exchange for a free beer or shot of whisky in his saloon. The Buckhorn collection of horns and antlers grew to be one of the largest and most unusual collections in the world. Albert's wife, Emile, got into it as well and began accepting rattlesnake rattles in exchange for a drink. She used them to create works of art still on display in the Buckhorn.

They've got fish from all seven seas, animals from every continent and strange animal oddities like the record-holding seventy-eight-point buck. They staged little shows there for people attending the museum and I would perform in them.

Busloads of people would come in there and I'd do poetry for them like "Wake Up Jacob" and "The Apricot Poodle Bold" and all of my old standbys. Some guys would sing and play music. Eventually a big brewery company bought it and made it even bigger and more commercial. After that it lost its atmosphere.

CORKY WRITES AND STARS IN HIS OWN PLAY WITH DANA

CORKY: There's a small poem in most of the cowboy books called "Wake Up Jacob." I've recited it often over the years and I recorded it on the *Bards and Pards* CD I did with David Longworth. There are all kinds of variations of it, but it goes something like this:

Wake up Jacob! Day's a-breakin'!
There's beans in the pot and the hoe-cakes a bakin'!
There's bacon in the pan and coffee in the pot,
Get up and git it, and git it while it's hot.
Roll out cowboy, dance me a jig,
Ya better come and get it, or I'm a feedin' it to the pig!

One day I sat down and turned "Wake Up Jacob" into a play. It's about this ranch cook, the people who worked on the ranch, and all the cowboys. It's a situation where the old Southern ranch owner dies, and his son sells the ranch to a Yankee northerner from New York City. A cultural clash ensues, resulting in the cowboys scheming to dissuade the new owner from staying around. The scheme backfires, and the new owner wins over the cowboys but decides to leave the ranch after all.

We staged it up in Kerrville, in the hill country of Texas. They had just built a beautiful new theatre there, the Cailloux Playhouse, that attracted a lot of big-name acts. Our play was the theatre's inaugural performance. I dedicated the play to all the working cowboys and ranchers I had known, and in particular to my grandfather, Luther Williams, and to Lester Dorsey, Bob Cohen and Maurice Tuck of Anahim Lake.

JEANINE: It turns out the main character in the play has a daughter he doesn't know about, so Corky's play is somewhat autobiographical. On the Kerrville stage Corky played the lead character and Dana played the lead female role as his niece, who at the end of the play is revealed to be his daughter.

DANA: Theatre runs in our family. Originally, my plan when I went to Austin was to pursue theatre. It's something my dad and I have in common. I only really worked with Dad onstage that one time when we did *Wake Up Jacob*. It was fantastic because he's so professional. Dad's character was Luther Lee Dorsey, the camp cook and ranch foreman. I played Kristen, his niece/daughter. Obviously, this is reflective of my dad's own life with his mystery child, so it was perfect.

We did it for four weeks on Thursday, Friday, Saturday and Sunday. I was living in Austin, so I'd drive two hours to Kerrville every Thursday after my workweek. Dad and I would get up in the morning and we would run lines together, hang out, then do the show. It was a really wonderful experience for me. I wouldn't trade it for anything.

Acting is the family business. It runs in our blood. I love it, and I hope to work back into it someday, doing it professionally. I had so much fun working with my dad. He was always the most prepared person on the set. Early in, last one out. He taught me that.

Dad was the kind of actor directors love because they could give him a big huge chunk of dialogue and he'd sit down and work and work and work and a day later come in and then bang, bang, bang. Perfect. He wouldn't miss a beat.

JEANINE: Dana was a very good actress. She won awards in high school and with the Williams Lake Studio Theatre. She went all the way to the provincials. She was talented, but needed to have something so she could make a living. She worked for a while waitressing and being a sitter, then somebody suggested she had a nice way with people. So she went to school for two years to become a massage therapist and she's done that off and on over the years.

END OF AN ERA

JEANINE: We saw the last of a certain kind of life in Anahim Lake. By the time we left, everything was already starting to change, but we got to see things like haying with horses and teams. We got to see the last of that generation of people, both Native and white, who were comfortable living in the land. Mickey Dorsey, Lester Dorsey and Pan Phillips all died within one year of each other, and a lot of the old Native people died at the same time. Every old-timer who died took their knowledge of how to do things and their knowledge of the country with them. It's gotten to where there isn't anyone left who knows-how to do it, even if you wanted to learn those skills.

In the business world they call that disappearance of skills, the loss of institutional knowledge. When you let too many of your older workers go, you lose a lot of inherent knowledge and know how that is often taken for granted. When I worked for the state government in Austin, Texas, it was a huge bureaucracy, but there were ways to get things done if you knew how. And you usually learned those skills from an older person who had worked there for a while. By the time those of us who were left had accumulated enough knowledge, we were on the way out too. Maybe it's a metaphor of our society. All the wise teachers are gone.

CORKY: I think what fools most people when they go out there to try ranching around Anahim Lake is how lush and green everything is. They see there's lots of water, and that especially impressed me since I was from the desert where we didn't have much water. You fall in love with the beauty of the country, but try and live in it and it's a different story.

If I were ever to buy a place again, I'd go out and live there summer and winter for a couple of years before ever settling on a place, because it can fool you. That green can fool you. It fooled me.

ANAHIM PEAK. THE CHILCOTIN GREENERY CAN BE DECEPTIVE.

RETURN TO CANADA

SAGE: Within a couple of years of each other Corky and Jeanine both returned to Canada after living in Texas for close to fifteen years. "I just had a wild hair up my ass to get up and come back to Canada," Corky says. "I always wanted to write something about what we were doing in Anahim Lake. It was time to do that. So I wanted to come back to the Chilcotin and make it go."

When Jeanine moved back to Williams Lake in 2007, Corky was already there. By that time their son, Little John, was a father. "Our grandson, Bryan, was one of the big motivators for me to come back to Canada and be closer to him," Jeanine says. "Also, I wanted to come home. To me Canada was home. It always has been. I felt like things were getting so haywire down in the States. The political situation was really upsetting. The wars, the street killings, the politics. I didn't learn my lesson the first time around, I guess, and it showed me again."

QUICK PISS SIS

SAGE: Corky had a little black and white dog named Sissy, and he gave her the nickname of Quick Piss Sis. Naturally Corky has a story about her.

A dog's life, being shorter than a human's, often serves to link the human owner to a certain time and place in memories. In Quick Piss Sis's case, her life linked Corky to the Anahim Lake backcountry and his transition from there to Rose Lake, Albion and eventually Texas. Both Corky and Jeanine took care of Quick Piss Sis as they went through their own changes; even as they were splitting up as a couple, they did what was best for the animal to ensure she was cared for.

CORKY: When it got real cold at Rose Lake or Muskeg, Quick Piss Sis would go to the door and we could see she wanted out. Then she'd back up and warm her hind end under the stove before running outside and pissing in a big loop, on three legs and one leg up just like a male dog. She never stopped, just pissed on the run all the way out and all the way back. Then she'd come right back into the house and go back under the stove to warm up. "I ain't moving." That's how she got her nickname Quick Piss Sis.

Sissy's mother, Rudy, was a purebred papillon. Jeanine called her Roo after the character in *Winnie-the-Pooh* but I called her Rudy. Rudy came from the Blackwater country. How she got there nobody has any idea. Somehow she got abandoned out there or left behind accidentally, and she showed up at Peter Alexis's place. She'd been living off scraps or anything she could find, and had become quite wild. We know for sure that she lived outside for two winters and it could have been three.

Peter Alexis would come and go back and forth from the Blackwater to Anahim Lake every couple of months to get groceries, and Rudy would follow Peter's horse and wagon, along with the rest of his outfit of dogs, horses and renegade hogs he had running loose in the Blackwater. You'd see them all come by.

When they got to Anahim Lake, Rudy would get beat up quite regularly by the bigger dogs that lived around town, and when she'd had enough, she'd head back home again on her own. A lot of times on her way home she'd head into our place. Fred's wife, Daisy, was Peter Alexis's daughter, so Rudy was quite familiar with Muskeg. We'd feed her and put food under the house for her because you couldn't get close to her. Some of Peter's grandkids had kind of beat up on her a little bit, and the other dogs would jump on her because she was small. Dogs always establish a pecking order, and she was at the bottom of the pile.

Different people would see her out on the side of the road, and I'd see her too, and I thought, "What in the hell is that dog doing out there in thirty-five or forty below, just catting along?" I know she killed a grouse one time in the spring when the grouse started mating and beating their wings on their chest to make that drum sound. She just went out and knocked off that wild chicken. That's how she made it, by hunting, but I never figured out how she avoided getting eaten by coyotes, bobcats, wolves or bear. She went down through the middle of it and survived it all.

One day she came into our place and we put food out for her in the spot she liked under the house, and she started to stay around. She'd come up there and eat and we finally got it so she'd come real close and we started petting her a little bit. She tamed right down, and she became Dana's dog. Dana loved that dog. When Fred was home Rudy stayed either at his place or my place. Then she got bred somehow, and that's where Sissy came from. Quick Piss Sis.

Rudy got bred at Rose Lake when Jeanine was living there with the kids. The only male dog I saw her with was a basset hound. Rudy got under the porch and the basset hound tore off a little piece of the front of the porch and made him a hole so he could go in there. For about four days you didn't dare get close to him. He'd bite the hell out of you. She was in heat and he was protecting her. I think she was his mate and that's where Quick Piss Sis came from.

JEANINE: Sissy was born at Rose Lake on Valentine's Day. I remember the day because Dana was so excited. The kids and I had already moved to Rose Lake when Roo came into our lives. I was out at the ranch for the summer with the kids when Dana came in and started talking about this little dog over at Fred's place. I caught a glimpse of this little critter with huge ears and a little tiny face. She had come in with Peter Alexis from the Blackwater and he had left her there at Fred's. She was just as wild as a little fox and nobody could get close to her. Dana absolutely went nuts for her, and she'd sit for hours with a little bit of food. Finally she got so she could touch her.

CORKY WITH GRANDSON BRYAN AND PUPPY FESTUS.

At the end of the summer Dana wanted to take Roo home to Rose Lake when it was time for us to go back in for school. Dana went working on Daisy, and finally Daisy, who was Peter Alexis's daughter, said she could take her.

I only had ranch dogs before and never a pet dog, but when we were ready to head back into town, Roo ran under the house and wouldn't come out. Fred and Daisy's son Garren crawled under the house and got Rudy, and she sat quavering in my lap all the way into town.

We didn't have a fenced yard at Rose Lake so I wondered how we were going to keep Rudy from running away, but it wasn't a problem. She moved into the house and was the best little dog. I fell in love with her. She was the cutest dog but we had no idea what her breeding was. We looked and looked in dog books, and finally there she was, a papillon. I'd never heard of this breed before.

Sissy's father was actually a cocker spaniel mix that lived down the road. He was a gypsy traveller. Once he bred Roo, we never saw him again.

Sissy and Roo were a pair all their lives. Corky and I both really loved the dogs, and even after we split up we both looked after them. We never had any problem doing whatever was best for them. Roo lived to be eighteen and Sissy was fifteen, and they went everywhere with Corky. They clocked more miles as he travelled back and forth from Texas. During the time Corky and I were together, the dogs stayed with me, but when I moved to Texas, Corky took them. Then I had them back again.

Roo was with Corky when she passed away, and Sissy was with me when I had to put her down.

CORKY: Both Quick Piss Sis and Rudy died of old age and we buried them in the same little graveyard at my brother Jaston's place overlooking this little lake. We keep the graveyard all cleaned up. They're Canadian dogs buried in Austin, Texas.

Jaston got wind of this lady who raised purebred papillons, and he bought Festus and gave him to me. He's about ten years old now and I brought him back to Canada with me when I moved back to Williams Lake. So Festus became a Canadian.

SAGE: Festus is Corky's constant companion these days. In the great small-dog tradition, Festus serves as the glue that stitches the fragments of Corky's life together. From the Chilcotin wilderness to Williams Lake to Texas and back again, the role of these small but mighty canines can't be overstated.

REFLECTIONS

JEANINE: I loved the ranching life. It was a hard life for women, mostly because men were always running off when you needed them, but it was also the kind of life where you learned to do things for yourself and be independent, and then you had the skills and everything you needed to take care of your family. It was a real sense of power that way. You weren't so connected and heavily dependent on the system, because you didn't have to be.

I learned I could make my way. I learned to cook meals with meat or without meat. I learned how to make things and how to run things. In the other world I never would have learned how to work well with a two-cycle motor, or so much about the cattle and the land. It was a different world, but it was a world I liked. Some women felt adrift in it, or they couldn't get their minds around it.

Author Diana French describes the Chilcotin Drummer in her book *The Road Runs West*. Everybody out in Anahim Lake marched to the beat of a different drum. You either heard that drumming or you didn't, and I always heard it. It was almost physical sometimes. When you crossed Sheep Creek Bridge over the Fraser River, that's where it began. That was always the cut-off. Sheep Creek was where you switched over from town to the country. I noticed that colloquialism from the beginning. People referred to whether you were in the country (in the Chilcotin) or whether you were out. You went out to go to work for a while, then you came back. It wasn't geographical so much as mental. You went out into "their" world to make some money, but then you came home.

Everybody in town knew where you were from because of the way you were dressed. We stood out. The women looked like they hadn't been into a beauty parlour. You could always tell the rigs from out there in the country, too. They were covered in mud and bashed up, because they'd been worked on in various ways just to survive the conditions. It made them look strange. Every once in a while I still see one around town, and I'll say, "Ha, that one's from the Chilcotin."

Betty Altmier, Goldie Reed's daughter, said something to me once that stood out. When I first came to the country, I was full of romantic wonder about everything. She laughed and said, "Well, I'll tell you what. Give it eight years, and you may feel different about it." You

CORKY IN 2012.

know, it was about eight years later that I realized I was going to have to go. I would have stayed with it longer, if things had been different. But I noticed that a lot of people would come and go after eight years.

During our time in Anahim Lake we got used to a lifestyle where we moved about to different places at different times. You learned to be content wherever you were. Strangely enough, I loved it. Once you give up your attachment to things, to stuff, it's easy to do. When we arrived at Corkscrew from California, we brought all our furniture with us. We had couches, chairs and beds, and I even brought my little upright piano. We left it all there at Corkscrew when we moved to Muskeg. It didn't seem that hard at the time. It didn't seem like we were losing anything. We were just going to a smaller place where we couldn't fit it in.

You didn't have to live in one place. You could live in this cabin or that cabin or you could build another one. The attachment to material things loses power over you. This can be upsetting to some people, but for me it was freedom.

DANA WILLIAMS: My dad has a streak in him that is so compassionate, so emotional and so sentimental, yet he has a part of him that is so cantankerous and unyielding. It's difficult to believe it can all exist in one person. When Dad was young, he wasn't the kind of kid who would lie down for anything. He would fight to the death. When he was three or four years old he would walk out of his yard and Grandmother would say, "Corky, where are you going?" And he would say, "I'm going down the road to pick a fight." That's just his personality.

My brother and I had to accept a long time ago that our dad was not like every other dad. He's not going to show up at birthday parties or graduations. He's not that kind of a dad. So we found other people to be close to as well.

But I love my father dearly. He is very gifted and talented and there's nobody else on this planet that I love more than him. If we're walking down the street together and we get to a stoplight, even at my age, he will still put his arm out to protect me.

Dad can tell a story like nobody's business, and you never have to wonder what he's thinking or how he feels about something. His life is a blessing to me. It always has been.

CORKY: I wouldn't trade my life in the Chilcotin for anything. But I'm too old to tackle it again. I don't have enough time to do it all over. There were sad times and there were also great times and wonderful people. There were a few assholes every now and then, but that's all part of it. I've still got friendships that will never die.

It sure taught me a lot about life and how to accept things. In 1971 it was a different time. There were gobs of people moving to the bush and the old-timers were still there, but they're gone now. It was a privilege meeting all those people and associating with them.

We did get to see the old way for a bit. The old-time way. They went out there with bare-ass nothing: a wagonload of groceries, a few saddle horses and a couple of plough horses and they headed for the boonies to make a living on the land where nobody else could, make the thing bloom a bit. It took a lot of hard work. It didn't happen by itself.

We had lots of trials and tribulations. Getting your tractor stuck showed how willing your neighbours were to help you. I got a chance to add my expertise to the land as well, with my rotovating and farming. I enjoyed that because I got to meet the Native people and hear their side of the story.

By the time I got to the country you could fly over the land and look at it. When the old-timers got there they couldn't do that. If they wanted to go look at a meadow in the park, they had to go there the best way they knew how. It might be under snow in June or July—they had no way of telling.

We had to experience both the good and the bad. Even a bad experience will teach you a lot. You learn not to mess with certain things because it's not going to work. You go out and look at big meadows and see the great hay potential, but there's ditches to put in and fences to build, and road access, and that's all hand work.

Big Fred, he knew just what to do. You couldn't have a better guy helping you. He was born and raised in that country and he was real smart. It was a privilege to have him around.

I'm too old to do it again and I wouldn't want to tackle it now. I want people to understand that was a tough row to hoe. A tough, tough road, as hard as I've ever taken. It's beautiful country, but hard to make a living in it. Having that experience makes you capable of taking on any kind of a project. And that's a big one right there. It's the root on the plant.

APRICOT POODLE BOLD

I had bought a ranch in the Canadian West
Up in the BC wilds.
I was drawn by the cowboy way of life,
I had loved it since a child.

But here there were bear, near everywhere,
There was moose and deer and fowl
And sometimes at night, just out of sight,
We'd hear the big boy howl.

Well, I soon fell in with some Indian friends
That I met in that big land,
And soon I knew some cowboys, too,
And they extended me their hand.

There was one I met, that I shan't forget,
And he taught me the open range,
Was a cowboy who in '52 had
Moved to this land so strange.

Now, he knew the land like the back of his hand.
He knew every creek and draw,
And he became my good friend, this horseman thin,
And his name was Wild-Man Bob.

Now a friend of Bob's, for doing a job,
Had offered him a gift.
Here, choose your dog from this catalogue.
It'll give your life a lift.

Well, he thought about that, as he pushed his hat,
And pondered on what to choose,
'Cause the dog came free, as Bob could see,
And he had nothing to lose.

So he picked him a dog from the catalogue
And waited for him to come.
He had chosen the poodle, a noble breed,
By grabs, this could be fun.

He was shipped from the States, in a little wee crate,
And what waddled out was a sight.
Out come this prize, with weeping eyes,
He would put his own mother to flight.

He walked right out on his little bowlegs
And he shook his bubble head,
And we knew right off that a Chihuahua dog
Had slept in his momma's bed.

He was supposed to be pure, but we knew damn sure
This just wasn't the case.
He had little pig ears—he'd panic the steers—
And, good God, what a face.

He had little cross-eyes and knobby knees
And a bit of a crooked jaw.
He was apricot-coloured, with a little red ass,
And that, folks, is what we saw.

"If that's a poodle, I'll turn into a noodle,"
I heard the wild man say.
"Good God, Ben, I've been skinned again.
Hell, I'm getting drunk today."

So we broke out the booze—what could we lose—
And we poured out one for the pooch.
To our surprise, right before our eyes,
He lapped up all that hooch.

"Hell, here's a friend, he'll fit right in,"
The wild man said to me.
"Let's teach him to ride, dad-burn their hide,"
And the poodle grinned with glee.

Now the apricot poodle was a hell of a dog
And he loved to ride and drink,
And if he didn't get his daily grog,
He'd raise one hell of a stink.

He'd howl and groan and piss and moan
Until we filled his cup.
With his little pink tongue a-singing a song,
He'd lap all that whisky up.

Then he'd give us a smile and sit back a while
And when the whisky took hold
He'd shake his hide—he was ready to ride—
This Apricot Poodle Bold.

Now, the apricot poodle rode standing up
With his front feet on Wild Bob's back.
We built him a saddle, just behind the cantle,
Made out of an old tow sack.

He learned to ride like the wind, with his new-found friends,
Just like a rodeo champ.
But this poodle dog demanded his grog
Whenever we might camp.

We would fix him some chuck and fill up his cup
And throw some more wood on the fire.
Then without a peep, he'd drift off to sleep,
And dream about being a sire.

He dreamed of canine girls, with perfume and curls,
That trot on Vancouver streets.
And he'd give all his pay, this very day,
If one he could really meet.

In mornings he jumped in the middle of rumps
And rousted us out in the fog.
We had to get up: it was time for chuck,
And also his morning grog.

Right after that we were under our hats
And the poodle beamed with pride.
He seemed to say, "Let's get on with the day,
Hey, boys, I'm ready to ride."

So we'd get saddled up and ready to track
And I'd put the dog up on his perch.
And with his claws in that sack, and his paws on Bob's back
The poodle would never lurch.

He'd stand there all day, 'cause for him it was play,
And he had no fear of a fall.
We crashed across rivers, spruce jungles and swamps
And the poodle was having a ball.

What more could we ask, as in the sun he would bask,
There was horses and grub and grog.
And he felt no pain as we rode through the rain
The Apricot Poodle Dog.

He had his true friends, and sometimes he got gin,
And once he drank some Dram-boo,
But his favourite of all, as we rode through the fall,
Was an ice-cold glass of Labatt's Blue.

He'd sit there and grin, as it ran down his chin,
And his whiskers were covered with foam.
And this little pink scamp, who rode like a champ,
Just loved his cowboy home.

We rode herd 'til fall, when we gathered them all,
Went home and got ready to ship.
And the Indians would grin at the crazy white men
And the horse with a dog on his hip.

Yeah, and the tourists would gape, like a big bunch of apes,
Whenever we got near,
'Cause on the fly, we'd often stop by
And water at the Old Frontier.

Well, late in the fall when the wind commenced to squall
And we began to wean,
There was slop and rain and hail and sleet
And some cows turned damned mean.

And what happened then, I'll tell you friends,
It happened just like that.
An old wild cow, with a horny brow,
Had knocked the poodle flat.

He hit with a thud, right in the mud,
And I beat the cow off with my hat.
And Bob ran in and grabbed up our friend
And got him the hell out of that.

But we knew with a look, as the top hand shook,
That his riding days were through
And with a wee little cough, he just seemed to slip off—
There was nothing that we could do.

Well, we thought all day about things we could say
And how to tell our wives.
What could we do about this canine true
Who had become such a part of our lives?

Well, we buried that dog with a bottle of grog.
He had worked for minimum wage,
And the feats will be told of this poodle bold
Into another age.

So, with a howl and a hoot and a whisky salute
We sent the poodle on his way.
And two cowboys cried around the fire that night
As we thought of a happier day.

After all these years, when friends draw near
And we speak of times gone by,
My bowlegged friend walks right on in—
He just refuses to die.

So whenever we gather to spin our tales
And remember days of old,
We toast our little pink friend, who rides in the wind
The Apricot Poodle Bold.

THE BALLAD OF THE PINE BEETLE

The Pine Beetle is a little black bug,
Where he comes from no one knows.
He's eaten up all the standing pine
And now he's gonna eat your clothes.
He's looking for a home boys, he's looking for a home.

The first time I saw the Pine Beetle
He was out on the Chilcotin plain,
Next time I saw the Pine Beetle
He was riding on a BC train.
He's looking for a home boys, he's looking for a home.

Next time I saw the Pine Beetle
He was only here and there.
Next time I saw the Pine Beetle
He was riding my old blue mare.
He's found him a home, boys. He's found him a home.

The logger says to the Beetle,
What makes your head so red?
The beetle says to the logger,
It's a wonder I ain't dead.
I'm looking for a home, boys. I'm looking for a home.

Well, the logger got mad at the Beetle
And sent him off in a balloon.
The Beetle says to the logger,
I'll have myself a honeymoon.
I'm looking for a home, boys. I'm looking for a home.

The logger grabbed the Beetle,
Stuck him in the red hot sand.
Beetle said to the logger,
I'll take it like a man.
He's looking for a home boys, he's looking for a home.

The logger took the Pine Beetle,
Put him way down in the ice.
Pine Beetle says to the logger
It's mighty cool and nice.
I got me a home boys, I got me a home.

The logger said to the banker,
We're in a hell of a fix,
Beetle has eaten the timber
And left us just the sticks.
We ain't got a home, boys. We ain't got a home.

The logger said to his Missus,
What do you think of that ?
Beetle's done made a nest
In your best Sunday hat.
Gonna have him a home, boys. Gonna have him a home.

If anyone should ask you
Who it was that wrote this song,
Tell 'em it was a broke-ass logger
With wore-out britches on.
Beetle's found a home, boys. Beetle's found a home.

COWBOYS AND BUSHRATS

There are many men whom you might call strange
That live past the peaks of the Coast Mountain Range.

I've seen their tracks in this Jack pine land
Across the spruce jungles and the meadow lands.

There's Billy Woods who fiddled left-hand
And Alex Graham the wild horseman.

There's old Dave Wilson from Sucker Lake,
A mountain man who could have lived on snake.

It's Tom Baptiste, an Indian man whose shot was true.
Just ask the enemy in World War II.

It's Sam Sulin, the old Medicine Man
Who could take more cold than we could ever understand.

My old pal Swede Johson, whose back was hunched,
There was a pile of cows that man had punched.

And old George Turner from Kleena Kleene,
They say he ran with the Daltons and could get damned mean.

It's Ogie Caposse who died in a strange manner,
They lied like hell when they said he tripped the hammer.

And Ol' Nicolai, there was no tougher man,
Killed by a whistlepunk in a big cedar stand.

There's Andy Holte whose land I've roamed,
For many years it has been my home.

And Roscoe Wilmeth, a scholarly man,
Who passed me his bottle and passed me his hand.

There's Morton Casperson, a great cowboy that,
See him right there, a-waving his hat.

Yes, there are many men whom you might call strange
That live past the peaks of the Coast Mountain Range.

Their marks are there in the trees and snow,
I feel their spirits in the wind and cold.

This poem is of cowboys and bushmen pure and true.
If you have never known any, I feel sorry for you.

I'VE GOT THIS HABIT, RABBIT

I've got this habit, Rabbit,
It's worse than dope or booze,
And it's the kind of feeling
That I never want to lose.

It's called a-punching cattle.
It sets my mind afloat
And it sure as hell is better
Than working on a boat.

Yeah, I've got this habit, Rabbit.
It's called a-raising cows.
I guess things could be worse,
I could be raising sows.

But I never was a hawg man,
There seems no place to ride,
And I can't see making footballs
From an acorn buster's hide.

Yeah, I've got this habit, Rabbit.
Them cows get in your blood.
I'd never make a doctor.
And at law I'd be a dud.

I'd never be a trucker,
It just ain't my cup of tea.
And working in an office
Don't allow you to be free.

Yeah, I've got this habit, Rabbit,
And I don't know what to do.
I couldn't be a law dawg
Or the guy who makes a stew.

I could never be a singer
Or the dude who plays the blues.
I couldn't be a store clerk
Or the man who does the news.

Yeah, I've got the habit, Rabbit.
I live on Indian Time
And what those people taught me
Has shorely blown my mind.

So I'll keep a-punching cattle
Like any insane bloke
And grin when I tell my banker
That both of us are broke.

GONE COWBOYS

To all the top hands gone over the rise,
We sense your presence and we feel your smiles.

To Francis Cassam, the medicine man,
And to Old One Eye and Lester and Pan.

To Eagle Lake Henry, who hated the law;
To a Texas cowboy, my own Grand-Pa.

To Morton Casperson from Norway Land.
To old Fred Garlic from the Rio Grande.

To Pancho Villa from Mexico,
To the Aussie down under and the southern gaucho

To all you cowboys who are riding on high,
I'll try to bring whisky when I make the long ride.

NOTE: FIRST POEM CORKY EVER WROTE
WRITTEN FOR IAN TYSON FOR THE WORLD EXPOSITION 1986
BY LUTHER CORKY WILLIAMS

INDEX

A

Abuntlet Lake 22, 132
Albion 178, 179, **180**, 181, 182, 185, 190, 196
Alexis Creek 152, 164-166
Alexis
    Cellia 141
    Minnie 45, 142
    Peter 45, 110, 113, 125, 141-143, 168, 196-198
Altmier, Betty 52, 57, 200
Anahim Lake 7-10, 13, 15, **17**, 19-22, 24, 25, 27-30, 32, 39, 40, 45-57, 73, 78, 79, 84, 89, 93, 95, 98, 100, 101, 107, 108, 121, 124, 128, 131-134, 143, 145, 151, 170, 194, 200, Cattlemen's Association 33, Stampede 30, 47, 169, 170, 172, 173
Anahim Peak 73, **74**, 112, **195**
Andruss, Steve (Squirrel Eater; Pink Jesus) 121, 122
Anne Stevenson Secondary School 80
Archibald, Doug 132
Armstrong 79
Ashby, Dr. Clifford 15, 16
Austin, Texas 181, 182, 184, 185, 190, 192, 194, 199

B

Baxter, Don 128
Baxter's Café 128
    Store 27
Beachcombers, The (CBC television series) 8, **177**, 178
Beaver Creek 103, 143
Bella Coola 29, 32, **63**, 87, 88, 113
    River 24
    Valley 87
Besbut'a (Anahim Peak, Obsidian Hill) 73
Betty Creek 60, 139
Bill Lehman Meadow 25, 54, 55, 90, 91, 103, 121, 122, 125, 138
Blackwater
    Country 48, 60, 65, 66, 68, 76, 77, 113, 125, 141, 173, 196, 197
    River 123, 142, 148, 169
Blaney Meadow 52
Bookmyer, Pete 107, 125
Bordertown (CTV television series) 8, 178, 181, 183
Brady, Texas 14, 15
Bryant, Alfred 151
Buckhorn Saloon 190, 191
Bulyan, Sammy 122
Burnt Biscuit 9, 98, 99

C

C2 Ranch 132
Cahoose

Andy 33, 62, 74, 113, 116, **118**
Corinne 80
Family **77**
Joe 76, 78, 113
Mack 160, 161
Madeline 160
Mary Joe 76
Cailloux Playhouse 192
Cariboo 179
Flats 125
Radio 169
Carrier Language 72, 78, 80
Carter, Wes 141, 168
Casperson, Morton 20, 22, 26, 31, 213, 216
Cassam
Francis 216
William 168
Castner, Brian 131
Chantyman, George **22**, 23-24, 45, 141
Charlotte Lake 149
Chilcotin 7, 9, 21, 24, 54, 69
River 24
Chiwid 21, 149
Christensen
Andy 25
Creek 71, 98
D'Arcy 8, 17, 26, 28, 65, 84, 121, 151
Clearwater Lake 149
Clements, Ardyth 185, **186**, 187
Cless Pocket Ranch 24, 25, 29, 35, 52, 54, 60, 96, 98, 99, 146
Cohen,
Bob 9, 19 ,20, 22, 32, 35, **36**, 75, 85, 131, 132, 139, 156, 160-163, 167, 171, 192
Patrick **132**, 160
Cold Camp 101
Cold Camp Lake 100
Cook, Charlie 113
Corkscrew
Creek 19, 24-26, 32, 33, 38-40, 45, 52
Ranch 8, 19-21, 24, 32, 33, 42, 51, 54, 55, 60, 70, 87, 139
Culberson County, Texas 12

D

Dean River 8, 22, 24, 25, 33, 48, 54, 73, 74, 76, 96, 132, 134, 140, 160
Air Services 171
Road 92
Dorsey
Lester 19, 21, 47, 49, 52, 54, 74, 82-84, 103-105, **106**, 107, 125, 134, **135**, **136**, 138, 145-148, 166, **167**, 168, 169, 192, 194, 216
Mickey 55, 134, **168**, 194
Douglas Lake Ranch 181
Drift, Johnny 177

E

Eliguk Lake 65, 66, 148
Elkins
Baptiste 21, 149
Charlene 62

Daisy 22, 62, 70-72, 74, 87, 88, 133, 134, 158, 160
Fred (Big Fred) 9, **22**, 24, 26-28, 32, 39, 53, 62-66, 70, **71**, 72-74, **83**, 84, **85**, 87, 88, 90-92, 94, 97, 98, 103-107, 111, 113, 116, 117, 136, 138, 140, 144, 146-149, 151, 152, 156, 160, 163, 196, 203
Garren 62, **71**, 199
John Lawrence 62, 133
Shammy 62
Engebretson, Fred 96
Entiako
Country 76, 98, 113
River 99-101
Escott, Wayne 152
Expo 86 (Vancouver World's Fair) 7, 8, 175-177

F

Ford Theatre, The (Washington DC) 190
Fraser River 24, 37, 171, 200
French, Diana 200
Fuller, Peter 76

G

Gano, Bernie and Swede 125
Gauthier, Ron 81
Giles, Dick 127, 128, **129**
Gladden
David 74, **75**
Frank 74, **75**
Joe 74, **75**
Gloomy Point 84, 127-129, 140

Goose Point 24
Goat Herder 113, 114
Graham, Roy 32, 50, 51, 162, 163, 171
Grand Lake, Colorado 13, 34
Gregg, Scotty 149

H

Hance
Carmen 79
Jason 79
Ray 79
Ryan 79
Susan 9, 45, 47, 76, 78-80
Harris, Les 181
Hermsen, Leo and Vivian 122
Hobson, Rich 135
Holte
Ada 54
Andy 54, 213
cabin 31, **61**, **71**
Gary 54
Larry 54
Mike 9, 25, 54
property 54, 55, 60, 154
Tommy 76
Home Ranch 77, 168, 169
Hope, BC 13, 38
Hump Creek 25

I

Ildash 25, 54, 55, 61, 75, 76, 122,145
Ilgatchuz Mountains 73, 76, 96,

112, 172
Irene Lake 76, 142, 148
Irwin, Donn 125
Itcha
    Lake 36
    Mountains 73, 112
Izatt, Gloria 80

J

Jack
    Gene 99, 101
    John 101, 113
    Maddie 101, 113, 139

K

Kamloops 138, 171, 172
Karran
    Ken 74, 121, 134, 139, 143
    Linda 32, 74, 121, 134
Kerrville, Texas 192
Kidwell, Katie 175
Kimsquit 24
King
    Annette 173
    Glen 172, 173
Kiston 47, 78
Kleena Kleene 51, 56, 125, 127,
    128, 149, 152, **156**, 212
Klinaklini River 24, 125
Kluskus 45
Knight Inlet 24
Knot Lake(s) 131, 149, 151
Kudra, Joel 57

L

Laidman Lake 65-67
La Jolla Playhouse, San Diego
    California 185
Lampert, Bill 60
Langley, Dana (Sunee Yuho) 7, 69
Leon
    Bella 9, 25, 110
    Dogan 62, 98, 110, 140, 146,
    147
    Edward 148
    Georgie 9, 25, 110
    Lenny 160
    Liza 110
    Sammy 148
Lessard Lake 25, 127, 138
Lessard Lake Ranch 54, 55, 70,
    73, 74, 82, **83**, 88, 132, 136,
    173
Lies From Lotus Land 8
Longworth, David (Davie) **180**,
    181, 191
Los Angeles, California 8-10, 15,
    16, 68, 186
Los Vegas 177
Louis Squinas Crossing 22
Lubbock, Texas 10, 13, 88

M

Mars
    Hazel 58, 59, 134
    Bob 58
Marshall
    Dodie 157, 185
    Gary 157

Majuba 101, 139
McDonough, Mike 9, 107, **108**, 125, **126**, 127, 128, **129**
Merritt 181
Moody, Ollie 9, 30, 32, **141**, 172, 173
Moose Lake 67
Morrison Meadow 34, 132
Moxon
    Cam 133
    Louise 133, 134
Muskeg Meadow (Ranch) 8, 25, 31, 33, 54, 55, 60, **61**, **62**, **66**, 70, **71**, 73, 74, **77**, 81-84, **86**, 87, 88, 91, 92, 96, 98, 99, 104, 110, **112**, 114, 116, **117**, 119, 121, 122, 124, 125, 127-129, 134, 136, 139, 140, 145, 147, 151, 154-165, 167, 172, 196, 202

N

Nazko 21, 45, 149
Nechako 76, 99
Nemiah Valley 122-124, 169
Newhall, California 15
Nimpo Lake 24, 99, 125, 149
Nukalow, Ollie 149, **150**, 151

O

Obsidian Hill (Besbut'a, Anahim Peak) 73
Okanagan 79
Ol' Yeller **90**, 91, 97, 103, **104**, **105**, 106, 107, 109, 124

100 Mile House 175
150 Mile House 155, 157, 185

P

Palmantier, Madeline 21, 149
Phoenix, Arizona 182, 184
Phillips
    Diana 77, 169
    Pan 65, 135, 141, 168, 169, 194
Precipice Valley 74, 151
Purjue, Voyne 123

Q

Quesnel 45

R

Rainbow
    Lake 66, 68, 76, 77, 134, 148
    Mountains 25, 73
Raylene (Corky's daughter) 182, **183**, **184**
Reed
    George 57, 95, 96, 109
    Goldie 57, 134, 200
Richardson, Gordon 76
Riske Creek 175
Rose Lake 90, 154, 155, 157, **158**, **159**, 166, 185, 196-199

S

Sager

Bryce 54, 60, 98
Duke 171
Sherry 54, 60, 98
Saint Joseph's Mission 78
Salmon River 24, 76, **77**, 78, 95, 96, 101, 102, 113, 127
San Antonio, Texas 185, 190
San Diego, California 186
Schurr, Jim 96
Seals, John **88**
Sears, Joe 190
Sheep Creek Bridge 200
Sill
    Aggie 45, 47
    Eddie 110
    Minnie 76, 78, **79**, 102
    Pat 47, 62, 76, 78, **79**, 102
    Thomas 47
Sister Suzanne 56, 57, 121
Skinner, Lilly (Chiwid) 149
Sleepy Hollow 141
Spencer Dickie's drugstore 166
Squinas
    Celistine 24
    Christine 45
    Domas 45
    Eustine 70
    Louie 25, 160, **161**
    Mac 73, 82, 87, 110, 136
    Thomas 19, 24, 45, 50
Stuart, Harold 39
Stump, Stanley 48
Sulin
    Billy 20, 38, 40, 41
    Dick 20, 40, 41, 42
    Sam 151

T

Tatla Lake 7, 36, 54
Taylor, Lee 74
Texas Tech University (Lubbock) 13, 14
Toney, Martin 102
Tuck, Maurice (Tucker) 55-57, 113, 140, 170, 192
Tyson, Ian 8, 175, 177

U

Ulkatcho Village **46**, 47, 76, 78, 102, 139
Uskisula 47, 78

V

Vancouver 7, 8, 56, 152, 163, 168, 172, 175, 177-179, 207
Vanderhoof 45, 143
van der Minne, Dr. Dirk 51, 56-59
Van Horn, Texas 10, 12, 182, 183
Vannoy
    Denise 68
    Murray 65
    Randy 67
Vaughan
    Floyd 67, 131, 152, 171
    Lora 173

W

West
    Charlie 43
    Jeanie 43

P.L. (Pierre) 43, 44
Wheeler Bottom 125
Wiersbitzky, Bernie (Burnt Biscuit) 9, 98, 99-101, 113, 114
William, Eugene 122, **123**
Williams
    Adele 88, 182
    Dana 9, 52, 58, **62**, 69, 82, 88, **90**, 112, 114, 134, **136**, 154, 155, **158**, **159**, 164, 165, 168, 181, 182, **184**, 192, 198, 202
    Jaston 8, 12, 181, 185, 190, 199
    Jim 10, **38**, **53**, 88
    John (Little John) 9, 10, **18**, 40, **41**, 42, 50, **53**, 56, 58, 62, 66, 68, **71**, 84, **88**, **90**, 122, **123**, 134, 145, 154, **156**, **157**, **158**, 177, 178, 195
    Kenneth 10
    Mary Dane 10
    Nina Kay 10
Williams Lake 7-9, 17, 30, 51, 52, 69, 75, 78, 87, 90, 92, 145, 154, 155, 158, 169, 171, 175, 177, 181, 185, 195, 199
Wilmeth
    Francie 20, 131, **132**, **133**, 134, 162, 163, 171
    Roscoe 132, 213
Woman Hill 66
Woodward, Woody 74, 135
Wright, Dick 13, 34

Y

Yates, Mike 157
Yuho, Sunee (Dana Langley) 69

Z

Zigler
    Brent 68
    Jake 68
    John 66, 68, 77, 93
    Sandy 66, 68, 77, 93, 134